RENEWALS 691-4574
DATE DUE

**WITHDRAWN
UTSA Libraries**

Benefits Assessment
The State of the Art

In addition to its French and Italian origins, the English word risk *has two other possible derivations which are interesting to contemplate: the Vulgar Latin word* resecum, *meaning danger, rock, risk at sea, and the Greek word* rhiza, *cliff, through the meaning "to sail round a cliff." Though unattested, these classical sources are appealing to the editors for the visual image they evoke: Ulysses and his men making their perilous voyage between Scylla and Charybdis in the Strait of Messina.*

Technology, Risk, and Society
An International Series in Risk Analysis

Editors:

Joshua Menkes, *National Science Foundation, Washington, D.C., U.S.A.*
Vincent T. Covello, *National Science Foundation, Washington, D.C., U.S.A.*
Jeryl Mumpower, *State University of New York, Albany, U.S.A.*
Stuart F. Spicker, *University of Connecticut, Farmington, U.S.A.*

International Advisory Board:

Berndt Brehmer, *Psykologiska Institutionen, Uppsala Universitet, Sweden*
Timothy O'Riordan, *School of Environmental Studies, University of East Anglia, United Kingdom*
Henrique Machado-Jorge, *S. and T. Resources Survey and Analysis Service, JNICT, Lisboa, Portugal*
Giancarlo Pinchera, *ENEA–Direzioni Centrale Studi, Roma, Italy*
Saburo Ikeda, *Institute of Socio-Economic Planning, University of Tsukuba, Japan.*
Francis Fagnani, *Centre d'Etude sur l'Evaluation de la Protection dans le Domaine Nucléaire (CEPN), Fontenay-aux-Roses, France*
Ian Burton, *Institute for Environmental Studies, University of Toronto, Canada*
B. Bowonder, *Center for Energy, Environment, and Technology, Administrative Staff College of India, Hyderabad, India*
Howard Kunreuther, *Department of Decision Sciences, The Wharton School, University of Pennsylvania, Philadelphia, U.S.A.*
Robert Kates, *Center for Technology, Environment, and Development, Clark University, Worcester, U.S.A.*
Ortwin Renn, *KFA, Programmgruppe Kernenergie und Umwelt, Julich, West Germany*
Herbert Paschen, *Abteilung für Angewandte Systemanalyse, Gesellschaft für Kernforschung GmbH, Karlsruhe, West Germany*
Charles Vlek, *Instituut voor Experimentele Psychologie der Rijksuniversiteit Groningen, The Netherlands*
Pieter Jan M. Stallen, *Centre for Technology and Policy Studies (TNO) Netherlands Organization for Applied Scientific Research, Apeldoorn, The Netherlands*
Kristin Shrader-Frechette, *Department of Philosophy, University of Florida, Gainesville, U.S.A.*

Benefits Assessment
The State of the Art

Edited by

Judith D. Bentkover, Vincent T. Covello,
and Jeryl Mumpower

D. Reidel Publishing Company

A MEMBER OF THE KLUWER ACADEMIC PUBLISHERS GROUP

Dordrecht / Boston / Lancaster / Tokyo

Library of Congress Cataloging-in-Publication Data
Main entry under title:

Benefits assessment.

    (Technology, risk, and society; v.1)
    Includes index.
    1. Cost effectiveness – Addresses, essays, lectures.
2. Policy sciences – Addresses, essays, lectures.
3. Public welfare – Cost effectiveness – Addresses,
essays, lectures.   I. Bentkover, Judith D.   II. Covello,
Vincent T.   III. Mumpower, Jeryl, 1949–   .
IV. Series.
HD47.4.B46   1985   361.6'1   85–19438
ISBN 90–277–2022–3

Published by D. Reidel Publishing Company,
P.O. Box 17, 3300 AA Dordrecht, Holland.

Sold and distributed in the U.S.A. and Canada
by Kluwer Academic Publishers,
190 Old Derby Street, Hingham, MA 02043, U.S.A.

In all other countries, sold and distributed
by Kluwer Academic Publishers Group,
P.O. Box 322, 3300 AH Dordrecht, Holland.

All Rights Reserved
© 1986 by D. Reidel Publishing Company.
No part of the material protected by this copyright notice may be reproduced or
utilized in any form or by any means, electronic or mechanical
including photocopying, recording or by any information storage and
retrieval system, without written permission from the copyright owner.

Printed in The Netherlands

Contents

Preface		vii
Acknowledgements		ix
Authors		xi
Abbreviations		xiii

1. The Role of Benefits Assessment in Public Policy Development 1
   by Judith D. Bentkover

   Historical overview of regulatory policy ❀ Historical overview of benefits assessment ❀ Valuation of health benefits ❀ Assessment of safety ❀ Identification of environmental benefits

2. A Conceptual Overview of the Foundations of Benefit-Cost Analysis 13
   by V. Kerry Smith

   Theoretical foundations ❀ The concept of benefits ❀ The concept of costs ❀ The role of time and uncertainty ❀ Benefit-cost analysis in comparison with other frameworks

3. Benefits Assessment of Regulatory Problems 35
   by Howard Kunreuther and Lisa Bendixen

   Considerations in evaluating benefits of regulation ❀ Benefits assessment within the policy process ❀ Five-step procedure in applying Executive Order 12291

4. Conceptual Framework for Regulatory Benefits Assessment 51
   by Baruch Fischhoff and Louis Anthony Cox, Jr.

   Scope and terms of the analysis ❀ Variables in benefits assessment ❀ Empirical assumptions in benefits assessment methods ❀ Choosing a method for benefits assessment ❀ Using benefits assessments

5. Theory of Regulatory Benefits Assessment: Econometric and
   Expressed Preference Approaches 85
   by Louis Anthony Cox, Jr.

   Use of individual preferences to define benefits ❀ The econometric approach ❀ Time, risk, and uncertainty ❀ A comparative evaluation of alternative approaches ❀ The social utility framework for benefits assessment

6. General Methods for Benefits Assessment 161
   by Ronald G. Cummings, Louis Anthony Cox, Jr., and A. Myrick Freeman III

   Direct monetary methods for evaluating public goods ❀ Indirect monetary methods for evaluating public goods

7. The Valuation of Risks to Life and Health: Guidelines for Policy Analysis 193
   by W. Kip Viscusi

   Willingness-to-pay principles ❀ Adjustment in the value of life: quantity, quality, and heterogeneity ❀ The human capital approach ❀ Valuations using market data ❀ Nonmarket valuations

8. Estimating the Benefits of Environmental Regulations 211
   by A. Myrick Freeman III

   The structure of a benefits estimation model ❀ Indirect methods for estimating individual use benefits ❀ Nonuse/intrinsic benefits ❀ Contingent choice

Reviewers 231

Index 233

Preface

In recent years there has been substantial interest in benefits assessment methods, especially as these methods are used to assess health, safety, and environmental issues. At least part of this interest can be traced to Executive Order 12291, issued by President Ronald Reagan in 1981. This Executive Order requires Federal agencies to perform benefits assessments of proposed major regulations and prohibits them from taking regulatory action unless potential benefits exceed potential costs to society.

Heightened interest in benefits assessment methods has in turn given rise to greater recognition of the inherent difficulties in performing such assessments. For example, many benefits that are intuitively felt to be most important are also among the most difficult to measure. It can be difficult to identify the full range of both benefits and costs. The choice of an appropriate discount rate for comparing benefits and costs over time is problematic. Even when benefits are quantifiable in principle and agreement can be reached on their valuation, required data may not be available. Thus considerable uncertainty is built into most benefit estimates, even when they are based on the best available data.

In light of the complexities and difficulties associated with the performance of a benefits assessment, this book reviews the current state of theoretical and methodological knowledge in the field. The review is extensive in that it covers over fifty years of research, theoretical development, and practice.

The book consists of eight chapters. Chapter 1, by Judith Bentkover, provides an overview of the volume and also includes a discussion of Federal guidelines for the use of benefits assessment methods in addressing health, safety, and environmental issues.

Chapter 2, by V. Kerry Smith, presents the conceptual framework underlying benefit-cost analysis. By identifying inherent assumptions and areas requiring judgment, the author provides guidelines for implementing the techniques and interpreting the results of a benefit-cost analysis.

Chapter 3, by Howard Kunreuther and Lisa Bendixen, describes the role of benefits assessment in health, safety, and environmental decision making. The chapter includes a discussion of detailed procedures for conducting regulatory impact analyses.

Chapter 4, by Baruch Fischhoff and Louis Anthony Cox, Jr., provides guidance in developing and implementing a model for specific benefits assessments. The authors discuss several topics, including (1) how to define the scope and terms of the analysis, (2) how individual and societal behavior influences the choice of appropriate methods and the interpretation of results, (3) how benefits assessment methods relate to expressed versus revealed preferences, (4) how to manage benefits assessments so that they generate useful, usable results, and (5) how and when to base decisions on benefits assessments.

Chapter 5, by Louis Anthony Cox, Jr., describes alternative approaches to benefits assessment. This chapter includes a comparison of the two major approaches to benefits assessment, the econometric approach and the social utility approach. Methodological tasks such as sample selection, aggregation of individual values, representation of preferences, and treatment of uncertainty are also discussed.

Chapter 6, by Ronald G. Cummings, Louis Anthony Cox, Jr., and A. Myrick Freeman III, describes a number of practical techniques of benefits assessment that are now being used or are being considered. For each technique the authors describe the principal strengths, weaknesses, inherent assumptions, and appropriate areas of application.

Chapter 7, by W. Kip Viscusi, discusses what is perhaps the most controversial topic in benefits assessment: the economic value of life and health. The author discusses procedures for measuring the value of life and health and notes a number of problems in implementing these procedures.

Chapter 8, by A. Myrick Freeman III, identifies and evaluates the economic models and available techniques for estimating the benefits that can flow from environmental regulations. The author focuses on the benefits of improved productivity, improved opportunities for recreation, and aesthetics. For each case there is an overview of benefits assessment problems and an identification of the data required to implement various benefits estimation methods.

Support for the study leading to this book was provided by the National Science Foundation's Division of Policy Research and Analysis through a grant to Arthur D. Little, Inc. Judith D. Bentkover, while on leave of absence from Arthur D. Little and at Harvard University, served as the project director. The Arthur D. Little, Incl. staff consisted of John Ehrenfeld, Lisa Bendixen, David Sakura, and Louis Anthony Cox, Jr. The National Science Foundation project staff consisted of Vincent Covello, Jeryl Mumpower, and Joshua Menkes.

All chapters in the book are original contributions and were commissioned, reviewed, and in many cases co-authored by members of the project staff. The opinions expressed by the authors are solely their own and do not necessarily represent the views of the National Science Foundation.

Acknowledgements

We would like to express our gratitude to the project steering committee for their substantial contributions to this volume. This committee consisted of Kenneth Arrow, Stanford University; Michael Baram, Boston University; Robert Dorfman, Harvard University; Ward Edwards, University of Southern California; and Aaron Wildavsky, University of California, Berkeley. Lester Lave, of the Carnegie-Mellon University, served as chairman of the steering committee and we are deeply indebted to him for his guidance and assistance. Special thanks are due to Joshua Menkes, Head of the Policy Sciences Section, National Science Foundation, for his expert advice and editorial contributions and to Diana Menkes for her professional editing of the manuscript. We would also like to thank Peter House, Director of the Division of Policy Research and Analysis, National Science Foundation, and Brian Mannix, Office of Management and Budget, Executive Office of the President, for their support of the project. Finally, we would like to thank the reviewers listed on page 231 for their helpful critiques of all or parts of this volume.

Authors

Lisa Bendixen
  Arthur D. Little, Inc.
  Cambridge, Massachusetts

Judith D. Bentkover
  Harvard University
  Cambridge, Massachusetts
  (on leave of absence from Arthur D. Little)

Louis Anthony Cox, Jr.
  Arthur D. Little, Inc.
  Cambridge, Massachusetts

Ronald G. Cummings
  University of New Mexico
  Albuquerque, New Mexico

Baruch Fischhoff
  Decision Research
  Eugene, Oregon

A. Myrick Freeman III
  Bowdoin College
  Brunswick, Maine

Howard Kunreuther
  University of Pennsylvania
  Philadelphia, Pennsylvania

V. Kerry Smith
  Vanderbilt University
  Nashville, Tennessee

W. Kip Viscusi
  Northwestern University
  Evanston, Illinois

Abbreviations

AI	additive(ly) independence or independent
BAT	best available technology
BCA	benefit-cost analysis
CEA	cost-effectiveness analysis
CPSC	Consumer Product Safety Commission
CRM	contingent ranking method
CV	contingent value
CVM	contingent valuation method
DI	difference independence or independent
E.O.	Executive Order
EPA	Environmental Protection Agency
EV	expected value
HPFM	household production function method
HPM	hedonic price method
IRR	internal rates of return
MANOVA	Multivariate ANalysis Of VAriance
MAU	multiattribute utility
MPI	mutual preferential independence
OMB	Office of Management and Budget
OSHA	Occupational Safety and Health Agency
OTA	Office of Technology Assessment
PM	petition method
PPI	potential Pareto improvement
RIA	regulatory impact analysis
SEB	subjective expected benefit
SEU	subjective expected utility
SIC	Standard Industrial Classification
SSM	site substitution method
TCM	travel cost method
UI	utility independence or independent
UIIGI	unanimous indifference implies group indifference
WDI	weakly difference independent
WTA	willingness to accept
WTP	willingness to pay

Chapter 1

The Role of Benefits Assessment in Public Policy Development

Judith D. Bentkover

A number of major Federal initiatives have been undertaken in the last two decades to promote more rational and less burdensome policy making. In general, such initiatives have involved systematic processes for reviewing both proposed and existing regulations and for issuing new regulations. These actions have placed a number of requirements on regulatory agencies regarding the documentation, quantification, explanation, and justification of their actions.

Federal agencies are now often required to perform benefits assessments of proposed regulations and are encouraged to refrain from taking regulatory action unless potential benefits outweigh the potential costs to society. Furthermore, ideally, regulatory objectives are chosen to maximize the net benefits to society.

Historical Overview of Regulatory Policy

Early Regulatory Requirements and Executive Order 12044

Early requirements for developing economic and environmental impact assessments were defined in Section 102 of the National Environmental Policy Act of 1969, in Executive Order (E.O.) 11949 (and the preceding E.O. 11821), and in the Office of Management and Budget (OMB) Directive A107 for inflationary impact statements. These regulations required that economic and environmental impact statements be developed and included in all recommendations or reports on proposals for legislation and other Federal actions.

On March 23, 1978, E.O. 12044 was issued, directing each Executive Branch agency to undertake several specific steps aimed at the improvement of existing and future regulations. This E.O. required:

1. Simplification of regulations to minimize cost and paperwork burdens on the public.

2. Reform of the regulatory development process to include publishing, at least semiannually, an agenda of significant regulations under development or review.

3. Agency-head approval.

4. Regulatory analysis exploring alternatives and their economic consequences, along with a detailed explanation of the basis for selection among alternatives.

5. Periodic review of existing regulations.

Provisions of Executive Order 12291

On February 17, 1981, E.O. 12291 was issued, calling for detailed regulatory analysis of all major rules. To be considered major, a regulation must be likely to result in an annual effect on the economy of $100 million or more; likely to result in a major increase in costs or profits for individual industries, levels of government, or geographic regions; or likely to result in significant adverse effects on competition, employment, investment, productivity, innovation, or on the ability of U.S.-based enterprises to compete with foreign-based enterprises in domestic or export markets [Section 1(b) (3)].

The effect of this order was threefold. First, it expanded the definition of "major rule" and so incorporated a greater number of regulations within its purview. Second, it granted the OMB authority to order that a rule not designated major by an agency head may be so designated. Third, it required that any set of related rules be considered together as a major rule. Executive Order 12291 established the following analysis requirements for major rules:

1. Administrative decisions shall be based on adequate information concerning the need for and consequences of proposed government actions.

2. Regulatory action shall not be undertaken unless the potential benefits to society from the regulation outweigh the potential costs to society.

3. Regulatory objectives shall be chosen to maximize the net benefits to society.

4. Among alternative approaches to any given regulatory objective, the alternative involving the least net cost to society shall be chosen.

5. Agencies are to set regulatory priorities with the aim of maximizing the aggregate net benefits to society, taking into account the condition of the particular industries affected by regulations, the condition of the national economy, and other regulatory actions contemplated for the future [Section 2 (a)–(e)].

To implement these analytic requirements, a new process for conducting regulatory impact analysis (RIA) was required. This process increases the time frame for promulgating regulations and specifies that an RIA must contain the following [Section 3(d) (1)–(5)]:

1. The Role of Benefits Assessment

(1) A description of the potential benefits of the rule, including any beneficial effects that cannot be quantified in monetary terms, and the identification of those likely to receive the benefits;

(2) A description of the potential costs of the rule, including any adverse effects that cannot be quantified in monetary terms, and the identification of those likely to bear the costs;

(3) A determination of the potential net benefits of the rule, including an evaluation of effects that cannot be quantified in monetary terms;

(4) A description of alternative approaches that could substantially achieve the same regulatory goal at lower cost, together with an analysis of this potential benefit and costs and a brief explanation of the legal reasons why such alternatives, if proposed, could not be adopted; and

(5) Unless covered by the description required under paragraph (4) of this subsection, an explanation of any legal reasons why this rule cannot be based on the requirements set forth in Section 2 of this Order.

This constitutes a considerable increase in the nature and amount of substantiation an OMB review requires to sustain regulatory actions. Moreover, Section 4, "Regulatory Review," requires that before approving any final rule, each agency shall [Section 4(a)–(b)]:

(a) Make a determination that the regulation is clearly within the authority delegated by law and consistent with congressional intent, and include in the Federal Register at the time of promulgation a memorandum of law supporting that determination.

(b) Make a determination that the factual conclusions upon which the rule is based have substantial support in the agency record, viewed as a whole, with full attention to public comments in general and the comments of persons directly affected by the rule in particular.

With respect to (b), weight must be given to general public comments and then "special" weight must be given to comments of persons directly affected by the rule. This appears to require a balancing of the concerns of both those incurring the costs and those receiving the benefits of the proposed regulations.

Only three types of regulations are exempt from these procedures: regulatory responses to an emergency situation; regulations for which these procedures would conflict with deadlines imposed by statute or judicial order; and such others as directed by the President's Task Force on Regulatory Relief (Section 8).

Executive Order 12291 has several implications. Briefly, it now requires:

1. Increased time requirements for the proposal, approval, and promulgation of regulations.
2. More rigorous demonstration of the *benefits* of the proposed actions, to the extent of weighing benefits against the societal cost.

3. Explicit analysis and selection of alternatives with the lowest societal cost.

4. More detailed and substantive analysis to support rulemaking.

Office of Management and Budget Guidance for Regulatory Impact Analysis

In order to assist agencies in satisfying the requirements of Section 2 of E.O. 12291, OMB issued regulatory impact analysis guidelines in 1981. These state that RIAs should be written to enable independent reviewers to make an informed judgment that the objectives of E.O. 12291 are satisfied. Specific guidelines for the development of RIAs state that the following be provided:

1. Statement of need for and consequences of the proposed regulatory action.

2. Examination of alternative approaches, including consequences of having no regulation and alternatives within the scope of the proposed action (e.g., less stringent permissible exposure levels, different effective dates, and alternative means of compliance).

3. Analysis of benefits and costs including estimates of present value expressed in constant dollars using an annual discount rate of 10 percent; specific type of benefits, when recieved and by whom; and the type of costs, when incurred and by whom.

4. Net benefit estimates including nonmonetary but quantifiable benefits, nonquantifiable benefits and costs, and cost effectiveness of various alternatives.

5. A rationale for choosing the proposed regulatory action (which should achieve the greatest net benefits to society).

6. Statutory authority.

The major implication of the OMB guidelines is that RIAs must not only explicitly consider nonregulatory alternatives but must also use risk and benefit assessments to link the market failure causing the problem to the proposed regulation.

Historical Overview of Benefits Assessment

Although common-sense principles of benefit-cost analysis (BCA) have prevailed for centuries, the application of formal BCA techniques is a twentieth-century phenomenon. One of the first applications occurred in 1902, when the River and Harbor Act directed the Corps of Engineers to assess the costs and benefits of all river and harbor projects. More widespread use occurred after the Flood Control Act of 1926, which explicitly required that only projects whose benefits exceeded their costs be submitted for congressional action. This act led to a proliferation of BCA as a basis for justifying various types of public investments and expenditures, including area development projects, defense projects, safety standard implementation, and health program development and implementation.

Today, BCA is widely used to assess public-sector resource allocation decisions where private-sector techniques, such as capital budgeting and return-on-investment analysis, are insufficient. There are several reasons for inadequacy of private-sector techniques in the public sector. For example, there are the issues of market failure related to public goods (where a service provided for one individual benefits all, as in the example of national defense); economies of scale (i.e., decreasing average costs as the size of a project increases, as in the case of a dam); and externalities (i.e., costs or benefits experienced by other than the immediate decision maker, such as the pollution of a community's water supply by a firm dumping waste material upstream).

Despite the conceptual simplicity and intuitive appeal of comparing benefits with costs, it is not a simple matter to identify, measure, and quantify benefits. The literature contains practical examples of benefits assessment in such areas as health, safety, and the environment, as well as expositions of such related economic concepts as consumers' surplus, spillovers, Pareto welfare issues, and discounting. Debates over the validity or credibility of these methods frequently surround efforts to develop "objective" assessments. Various efforts have been made to address these problems. In fact, a Public Health Service report recommended the establishment of uniform benefits assessment guidelines.

The current requirements for regulatory analysis described above make these issues important in a very immediate sense to a wide range of Federal agencies. Executive Order 12291 and the OMB guidance document are silent on how to measure social costs and benefits of a proposed regulation, leaving case-specific detail to the Federal agencies. In addition, each agency must not only comply with the E.O.s, OMB directives, and statutory laws when issuing regulations, each one must be mindful of pertinent Federal judicial decisions construing the authority and applicability of the enabling legislation under which the regulations are issued. As a result, each agency must deal with a number of substantive issues which affect its ability to analyze regulatory impacts, including development and use of state-of-the-art methodologies needed for the analysis.

As discussed below, because of the absence of specific direction regarding the use of particular benefits assessment methods, Federal regulatory agencies have developed their own technical expertise to address the complexities of health, safety, and environmental policy formulation. Each agency has had access to different data, has faced a different set of events (i.e., different combinations of frequency and cost), has been subject to pressures from diverse and changing interest groups, and has had experts in different academic disciplines. Thus, the valuation of benefits varies and often the same benefit is valued differently depending on particular agency constraints and practices.

Valuation of Health Benefits

Within a single decade, society's principal health system goal has shifted from improving access to care to controlling the rapidly inflating costs of care. The key health policy question today is how to contain health care costs without sacrificing the desired benefits of improved health. The theoretical solution to this problem would consist of identifying and eliminating "unnecessary" services. As time passes, increasing numbers of procedures and medical devices are being scrutinized more closely as the previous enthusiasm for more is being displaced by attempts to encourage less.

Although the health care cost inflation problem has been partially caused by the advancing state of medical technology, another major contributor has been the growth in third-party financial liability for health care services. Presently, for example, a physician can order expensive laboratory tests that have a low probability of improving a diagnosis and will not impose any financial burden on the patient. On a societal level, technologies that otherwise might not have been adopted at all have spread rapidly. In short, the market's ability to ration and efficiently allocate health care resources has deteriorated to a virtually nonexistent level. In this environment, benefit-cost analysis has been regarded as a means of providing guidance for society's allocation of resources in order to maximize society's health status.

In 1980 the Office of Technology Assessment (OTA) published an extensive bibliography of benefit-cost analyses and cost-effectiveness analyses (CEA) containing approximately 600 references, mostly from the years 1966 through 1978. It is interesting to note from the OTA analysis that growth in the number of health care related BCA/CEA articles has greatly surpassed the increase in the overall medical literature. This health care BCA/CEA literature has focused on a number of issues and policies, including:

1. Specific disease treatment and screening programs (e.g., cardiovascular disease, cancer, mental illness, drug abuse, alcoholism, and renal disease).
2. Health problem prevention (e.g., birth defects and communicable diseases).
3. Specific procedures (e.g., radical mastectomy, tonsillectomy, chol-

ecystectomy, herniorrhaphy, appendectomy, joint replacement, and hysterectomy).
4. Particular health care activities (e.g., screening, diagnosis, and prevention).

Benefit-cost analysis also has been used to analyze a set of alternative ways in which a given health or medical care service can be produced or distributed (e.g., alternative ways of reducing motor vehicle deaths and injuries).

In general, as Prest and Turvey (1965) point out, the major purpose of health programs is to save and reduce illness; in this regard, there is considerable overlap with other activities such as flood prevention and road improvement measures. As these authors further note, some of the differences in benefits assessment methodologies result from data availability rather than theoretical and methodological considerations.

Certain methodological issues arise often in the health field. For example, some studies compare the efficiency of life-enhancing programs by valuing limited-activity days at some fraction of healthy days. Throughout this literature there is no consensus on weighting schemes for alternative states of health. Another benefits assessment issue concerns the clear identification of the isolated treatment outcome. In particular, the attribution of a benefit to a single component of a treatment regimen is often difficult. A final example is the assessment of benefits accruing from regulations that affect hospital and physician reimbursement. These benefits are inherently more difficult to assess than regulations that require identifiable changes in the physical environment of health care institutions. As a result, it is necessary to develop assumptions about how hospitals and physicians will react to changes in their fiscal environment. This in turn requires an understanding of how physicians value leisure *vis-à-vis* earnings.

While, by itself, BCA often cannot provide sufficient justification for policy makers to implement a particular course of action, there is evidence that such studies have played an important role in public health policy determination. A well-known example of the use of BCA in health policy formulation concerns family planning and birth control. After receiving reported findings of benefits exceeding costs by a hundredfold, U.S. Agency for International Development funds to assist developing countries were expanded in order to implement such programs.

A significant constraint on the use of BCA has been the relative inadequacy of existing methods to deal with qualitative issues. For example, changes in the quality of life (such as reduced levels of pain, discomfort, and grief) are particularly difficult to assess. Such problems are not limited to the health area; indeed, many pertain to analysis of safety and environmental issues.

Assessment of Safety

Many applications of BCA tend to emphasize optimal allocation of limited resources among competing productive ends. In the area of safety, however, the concern is more with events having low probability but potentially severe consequences and with the adequacy of safety measures for such events. Benefits assessments in safety studies primarily focus on the avoidance of risk and the identification of cost-effective risk-reduction measures.

The consequences of risk-reduction measures might include:

1. Reduced chances of the release of hazardous material and of the destruction of property.

2. Reduced quantities of hazardous materials released if a release does occur.

3. Decreased numbers of mishaps involving multiple fatalities or injuries.

4. Decreased expected numbers of fatalities or injuries per mishap.

5. Decreased property damage or production losses.

6. Better distribution of risks across a particular group of people.

7. Reduced insurance premiums and avoidance costs.

Clearly, other benefits also could be considered, but the above items are not mutually exclusive. One of the major problems in this area is the selection of a set of benefit measures which is complete without being redundant. Another major issue is the value of safety and how the value of life changes when multiple-fatality incidents are considered. Difficulties encountered in safety-related benefits assessments include the fact that large, complex projects may not offer benefits (such as increased employment) to the same people who are at risk. Other difficulties include assuring that the assessment methodology is not overly sensitive to minor changes in the probabilities of the events of concern and also determining means to handle multiple-fatality/injury versus single-fatality/injury events.

In the case of the higher-probability single fatality, injury, or disability, which may occur in occupational settings, a possible definition for benefit is avoided cost. The avoided cost may be economic or may be more abstract (e.g., avoided pain from injury). Direct economic benefits would include the avoided treatment costs associated with prevented injuries, while indirect economic benefits would include avoided loss of output to the economy (foregone earnings) resulting from premature death, permanent partial disability, temporary disability, and absenteeism.

Benefits assessments are now being performed much more routinely as an integral part of safety studies, but there is no agreed-upon methodology for identifying or quantifying benefits. Moreover, the use of the results of such

1. The Role of Benefits Assessment

benefit analyses varies widely. That level of risk which is deemed acceptable is either not known quantitatively or varies depending on the decision maker or regulatory body. Hence, the overall reduction in risk which must or should be achieved is fairly arbitrary.

The most significant value of BCA for low-probability events (typical of the chemical, petrochemical, nuclear, and transportation industries) has been in comparing alternate sites, technologies, or designs. Modifications to existing facilities also lend themselves to the application of benefit-cost techniques.

The first step in estimating the benefits of safety regulation is to estimate the reduction or avoidance in injuries, illnesses, and fatalities that the standard will produce. Safety outcomes can be estimated from data on preregulation injury rates and causes (e.g., from Consumer Product Safety Council data) with (possibly judgmental) consideration of the effect a regulation will have in changing these patterns. These avoided injuries themselves represent an intangible benefit. The direct economic benefit of avoided treatment costs can be estimated using data available from the medical community and health insurers. The indirect benefit of avoided foregone earnings cost includes both the discounted present value of earnings foregone due to an interval of permanent disability and the present value of any earnings reduction related to partial disability. Injuries also may entail a number of noneconomic costs to the individual and to others, such as pain and anguish, diminution of capacity to enjoy life, and loss of affection. In addition, anxiety associated with the risk of exposure to injury may be considered a cost. Of course, such costs involve subjective valuation and are not easily identified. The values attached to various kinds of disability by workers' compensation insurance costs may be used to give a lower bound for the amount of such costs. However, problems of confounding with *ex ante* risk attitudes and the inability to compensate for certain irreversible damages *ex post* clearly inhibit simple interpretation.

Identification of Environmental Benefits

Environmental regulations create a wide variety of benefits. The primary effects are improvements in the quality of air, water, and land, including reduction of the amount of pollutants or unnatural substances present. Other effects include direct changes in physical attributes such as visibility, odor, and taste.

Changes in quality per se can be construed directly as aspects of benefits. However, in the context of E.O. 12291 it is important to relate these direct effects to a set of benefits perceived by human beings. One conventional categorization defines effects as those directly related to human health and those more broadly related to ecological systems or overall welfare.

Improvements in air and water quality lead to reduced illness and death.

Techniques of health risk assessment use toxicological relationships or epidemiological correlatons to convert changes in environmental quality to equivalent changes in morbidity or mortality. These changes are in turn converted to equivalent economic values. The methods used to convert health data into economic equivalents are the same for changes induced by environmental regulations as for changes induced by health and safety regulations.

Changes in environmental quality affect the state of ecological systems as well as of mankind. Improvements in air and water quality would be expected to lead to improvements in agricultural, forestry, and fishery production. The primary means of addressing benefits in these categories is similar to that used for human health effects. Correlations between production and environmental quality using essentially epidemiological techniques have been developed, particularly in relating agricultural productivity to air quality. Alternatively, one can study the specific dose-response relationships between individual species and measures of the air and water quality. For these productivity effects, monetary benefits can be estimated directly from market data.

Air and water quality changes also affect the recreational uses of ecosystems, that is, such activities as fishing and hunting. Since recreation is not simply traded in the marketplace like a commodity, indirect methods are required to evaluate benefits. There is a substantial body of methodology useful in this area, much of it developed in the process of justifying public investment in recreational facilities.

In addition, the quality of the environment has a direct impact on the stablity and diversity of the ecosystem. The existence of such statutes as the Endangered Species Act provides evidence that man values preservation of species and ecological diversity, although the art of converting those values into economic terms is relatively undeveloped. A number of efforts are underway to monetize ecological damage caused by the release of pollutants into the environment, but benefits of maintaining ecological stability and diversity are in general more often expressed directly rather than converted into monetary equivalents.

Finally, there are a number of environmental welfare benefits induced by physical changes without regard to human or ecological organisms. These generally include such categories as materials damages, soiling, direct effects on production, weather and climate modification, odor, visibility, and visual aesthetics. As in the case of recreation, most of these effects create indirect benefits, that is, benefits not directly reflected in the marketplace. Methodologies based on either property values or wage levels (hedonic approaches) or preferences expressed directly in questionnaires or through interviews (contingent valuation methods) can sometimes be used to assign values to a set of benefits perceived by humans. Such approaches have been used recently by the Environmental Protection Agency in

1. The Role of Benefits Assessment

assessing the benefits of potential changes in air quality standards for particulate matter.

Although theoretical means have been developed and discussed for almost all of the benefits categories described above, the application of these methodologies is fraught with difficulties. A high degree of uncertainty is involved in assessing benefits, and many of the relationships are merely statistical in nature. Heroic assumptions often must be made to extrapolate the results from case studies and laboratory work to estimates of national benefits. Great care must be taken to avoid double counting when using different methodologies simultaneously, since these methodologies may each include values for the same set of benefits.

Conclusions

In the policy-making arenas concerning health, safety, and the environment, some of the benefits that are intuitively felt to be most important are also among the most difficult to measure. They are often not readily quantified (e.g., improved quality of life) and even when units of benefit can be defined, their value in terms of dollars or other standard measures is a matter of subjective judgment. Even in the case of quantifiable benefits, problems arise when a value is attached not only to the benefit perceived by each individual but also to an equitable distribution of the benefit across a population group.

In addition to the major issue of qualitative benefits, there are a number of other difficulties in the application of benefits assessment. One such difficulty concerns the assumption that the economic environment for a proposed action (prices or quantities of the supply of factors of production or the pattern of demand) will not itself be changed by the action. For large-scale projects or actions with far-reaching consequences, this assumption does not hold. As Dunlop states in a review of BCA in health care, "unfortunately the basic microstructural assumption underlying the benefit-cost analytical framework makes it particularly vulnerable when applied to health problems with a potential for a multiplicity of interactions with other societal institutions, not the least of which being a demographic structure" (1975:134). For example, if malaria were to be eradicated in developing countries, there would be a major societal impact realized during the subsequent 20 to 30 years. Only with the appropriate combination of macro and micro analytical frameworks can appropriate policy recommendations be developed.

A number of other problems are associated with the methodology used to assess benefits. These include difficulties in identifying the full range of both benefits and costs (including those accruing to other than the sponsor and intended beneficiaries); differentiating benefits from costs; choosing an appropriate discount rate (if any) for comparing costs and benefits over time; selecting the appropriate criterion for comparing benefits (e.g., net present value or benefit-cost difference); and handling multiattributed and intermediate outcomes. In addition, the amount and type of sensitivity analysis conducted are important elements in a benefits assessment.

In discussing problems associated with benefits assessment, Dunlop and others point out that the concept provides a useful framework for thinking about a proposed action or comparing alternatives even when this cannot be done in a wholly quantified way. However, without better resolution of the methodological problems, proponents of an action may tend to assign unduly high values to those benefits which are important but hard to measure, while opponents may tend to overemphasize the fact that the benefits which are easiest to quantify and value are relatively small.

Harvard University and Arthur D. Little, Inc.

References

Dunlop, D. W. 1975. 'Benefit-Cost Analysis: A Review of Its Applicability in Policy Analysis.' *Social Science and Medicine* 9 (March): 133–138.

Office of Technology Assessment. 1980. *The Implications of Cost-Effective Analysis of Medical Technology, Background Paper 1: Methodological Issues and Literature Review.* Washington, D.C.: U.S. GPO.

Prest, A. P. and R. Turvey. 1965. 'Cost Benefit Analysis: A Survey.' *Economic Journal* (December): 683–735.

Chapter 2

A Conceptual Overview of the Foundations of Benefit-Cost Analysis

V. Kerry Smith

Benefit-cost analysis has, in many policy contexts, been considered any analytical method that enumerates the advantages and disadvantages of alternative actions.[1] When interpreted in strict economic terms, it is a pragmatic realization of the theory of welfare economics, providing a specific organizing framework and a set of procedures to summarize information and display the tradeoffs associated with these actions – generally in monetary terms. While a number of past descriptions of the analysis have been so general as to permit many different types of methods for evaluating actions under the heading of benefit cost analysis (BCA), this chapter focuses on the narrowest of these interpretations – the strict economic conception. In this setting, BCA judges actions based on an efficiency criterion. Positive aggregate net benefits imply the prospects for an improvement in the resource allocation (over the status quo). Anything else does not!

Economic efficiency requires that resources be allocated to their highest valued uses. Moreover, in Western market-oriented economic analyses of these decisions, the values are assumed to be derived based on the principle of consumer sovereignty. That is, individuals are considered to be their own best judges of the values derived from the goods and services being provided. In general, the analysis is based on the premise that the action under consideration affects a small fraction of the resource allocation decisions of economic agents (i.e., households and firms) so that relative prices do not change. The public sector is viewed as intervening in the resource allocations to "correct" a departure from efficiency. Clearly, this last assumption is an extreme simplification. Indeed, in some cases, the very policy conditions requiring that a benefit-cost analysis be conducted themselves imply a violation of this assumption. For example, Executive Order 12291 requires that a benefit-cost analysis be conducted for *major* regulations – having an annual impact of $100 million or more on the economy. These would certainly have the potential for affecting relative prices. Consequently, what is important in practice is the extent to which the violations of such assumptions affect the plausibility of the estimates provided by the partial equilibrium methods of analysis that generally form the basis for deriving the benefit-cost results.

With this general background, it is now possible to discuss in somewhat more abstract terms the theory and practice of benefit-cost analysis. The following section provides such an overview, considering both the specific conceptual foundations for BCA and the concepts used in implementing that theory. The next section compares BCA with seven other decision frameworks, and the last discusses its role in the policy process.

Theoretical Foundations

Conventional economic analyses of static resource allocation decisions define an efficient allocation of resources in marginal terms with the Pareto criteria. These conditions are based on the principle that there will be an evaluation of all possible changes to the allocation of each resource. As part of this evaluation, changes in the use of factor inputs and in the allocation of outputs will be considered. In each case changes will be accepted provided they improve the welfare of at least one person without reducing the well-being of anyone else. When there are no more of these opportunities, the allocation is termed Pareto efficient.[2] An economy with competitive markets governing the exchange of all goods and services (including inputs), no externalities, and no other impediments to the allocation of resources will lead to a Pareto-efficient allocation of resources. Given the distribution of income and wealth, these conditions assure that resources are allocated to their highest valued uses.

In formal terms, Pareto efficiency implies that the marginal benefit realized from having access to a good will exactly equal the marginal cost of producing it. For private commodities this assures equalization of these marginal benefits (in real terms) across individuals for each commodity and across productive uses for each factor input. For public goods (where consumption by one individual does not diminish the amount available to others), the sum of marginal benefits (across the individual consumers) will equal the marginal costs of provision.

Since Pareto efficiency is a theoretical ideal, three issues must be considered in the transition from this ideal to its interpretation in the benefit-cost "test" for evaluating actions. The first concerns the evaluation of movements toward an efficient allocation. That is, we can consider the Pareto criterion as a basis for comparing alternative states, in that we judge one state superior to another when at least one individual's well-being can be improved without making anyone else worse off. Such an action is one of the set of accepted reallocations that together define a Pareto-efficient point. Each action is usually termed a Pareto-superior state because it is considered a movement toward efficiency. When all such moves have been taken, the Pareto-efficient allocation is realized. As a rule, benefit-cost analysis has been used in these types of decisions, which are, at best, movements toward an efficient allocation but do not identify the efficient choice.

In nearly all cases there are individuals who gain and those who lose,

2. The Foundations of Benefit-Cost Analysis

without corresponding compensation from the gainers to the losers. This is the second pragmatic adjustment accepted in moving from the ideal to the real-world implementation. BCA accepts an aggregate measure of the net benefits to judge actions. A finding of positive net benefits implies an improvement in the resource allocation. Therefore the analysis accepts the Kaldor-Hicks compensation principle. Potential (as opposed to actual) compensation is sufficient for actions to be judged improvements. Thus BCA relies on the criteria of a potential Pareto improvement (PPI) in judging changes in the resource allocation.[3]

The third issue relevant to the transition from the Pareto criterion to the benefit-cost condition arises from the marginal conditions that characterize a Pareto change and the incremental (i.e., often fairly large discrete changes) associated with the benefit-cost comparison. BCA focuses on the difference between increments to aggregate (across individuals) benefits and to costs as a result of the action under consideration. The existence of positive net benefits confirms the inefficiency of the base point or frame of reference (see Bradford 1970).

Fisher and Smith (1982) have proposed a simple diagram to illustrate the distinction. We could characterize a Pareto-efficient solution in terms of the aggregate benefits and costs for some desired service flow A, holding all else constant. Figure 2–1 plots these aggregate functions as $B(A)$ and $C(A)$, respectively. In terms of this one service flow, the Pareto-efficient conditions would require that the level of A be selected so that the aggregate marginal benefits of A were equal to the aggregate marginal costs, as at point A^*. As the figure illustrates, the vertical distance between $B(A)$

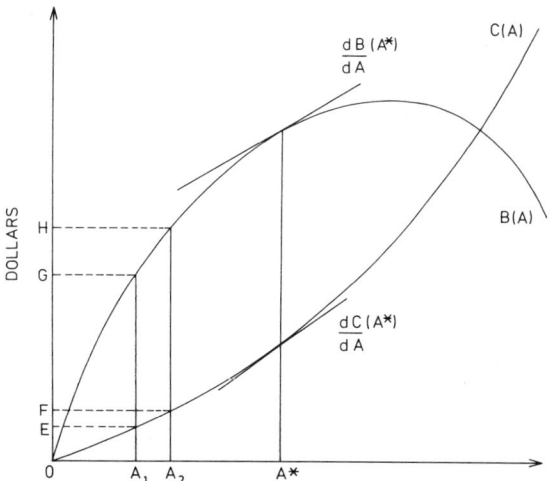

Fig. 2–1. Comparison of Pareto efficiency and benefit-cost criteria, from Fisher and Smith (1982).

and $C(A)$ is largest at A^*. By constrast, benefit-cost analysis considers whether the change from a given state will be desirable – for example, from A_1 to A_2. Such a comparison for the benefit-cost "test" examines the incremental aggregate benefits GH in comparison with the incremental aggregate costs FE. If the difference is positive, then action moves the allocation in the direction of an efficient allocation (i.e., toward A^*). This will be true regardless of the position of A_1 in relationship to A^* (i.e., it can be above or below A^*, provided the aggregate benefit and cost functions are well behaved). The adaptation of the theory of welfare economics to the benefit-cost criterion represents a substantial oversimplification. Nonetheless, the actual practice of using BCA in the evaluation of specific actions is not a routine translation of the theory just outlined. Consequently, we will describe some of the features of that practice in order to highlight several of the important assumptions that underlie its use.

The Concept of Benefits

Benefit-cost analysis maintains that consumers' values should be the basis for measures of the benefits of an action. In defining them, economists have generally used an individual's *willingness to pay* for the good or service provided by the proposed action. When the commodities involved are exchanged in perfectly competitive markets, then the market price measures the willingness to pay for the last unit consumed, and the area under the demand curve the total willingness to pay (with given income) for all that is purchased. Thus, for commodities which are exchanged on such markets, benefit measurement is reasonably straightforward. What is needed is to consider how the action being contemplated affects the availability of the commodity. Most of the economic theory of welfare measurement has maintained that actions affect the prices of the commodities involved. Thus, evaluating the benefits of an action that reduced prices would amount to estimating how the individual would value the price decline.

While this task seems clear-cut within those definitions, reviewing it in more detail is desirable because the process does identify issues that are important to the application of benefits analysis. First, using an ordinary (i.e., Marshallian) demand function, one measure of the value of a price change would be the increment to an individual's consumer surplus. The consumer surplus is simply the excess of the individual's willingness to pay over the expenditures he must actually make for a good or service at a fixed price; that is, at price OB, BAC in Figure 2–2. Thus, using Figure 2–2, for a price decline from OB to OE, the increment would be $BCGE$. This welfare measure is both common and controversial. The reasons for its use are direct: until recently it has been regarded as the easiest to evaluate and as having the greatest chance of being estimated based on the observed actions of households.[4] Unfortunately, it is not an ideal welfare measure, because

2. The Foundations of Benefit-Cost Analysis

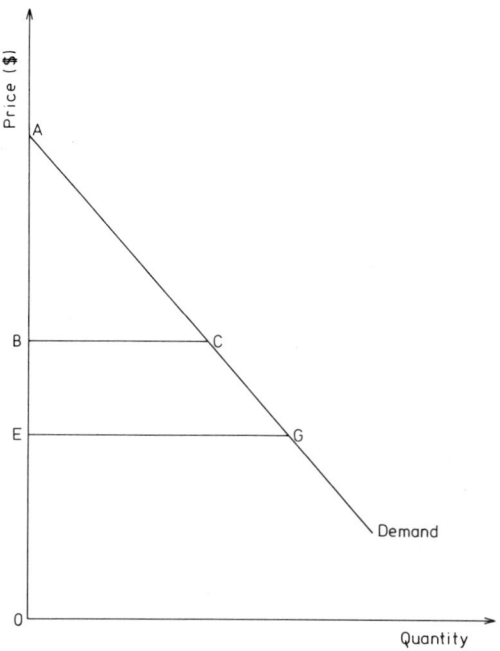

Fig. 2–2. Illustration of consumer surplus change associated with price increase.

along an ordinary demand curve, *income*, and not an individual's *well-being* or utility, is held constant. To evaluate how an individual would value a price change, we need to hold his well-being constant.

The *Hicksian* measures of welfare change using Hicks-compensated demand curves, provide such concepts, describing either how much an individual would be willing to pay for a price reduction, holding utility constant at the level realized with the initial prices and income (compensating variation), or the income change equivalent to the price change to realize the new utility level but maintain the old prices (equivalent variation). A simple means for illustrating the potential difference between these measures is to consider the valuation of a price reduction measured with each surplus concept. The Marshallian measure of the consumer surplus would be the increment to the area under an ordinary demand function (where income is held constant) as a result of the price change. Hicksian surplus measures can be defined in an analogous manner using demand functions constructed to hold the individual's well-being (i.e., total utility) constant at either the pre-price reduction level (for the compensating measure) or the post-price reduction level (for the equivalent measure). All three demand functions are constructed in Figure 2–3. The Marshallian measure falls between the compensating and equivalent variation measures. It overstates the former by *ADC* and understates the latter by *ABC*. These

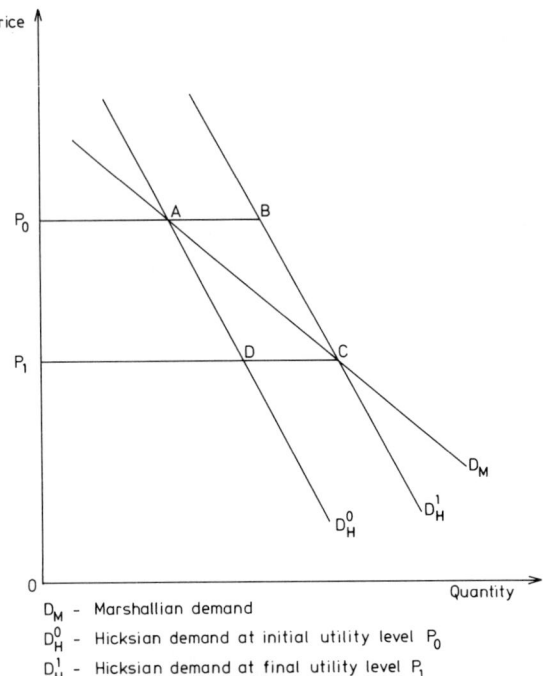

Fig. 2–3. Illustration of relationship between Marshallian and Hicksian surplus measures.

triangular areas provide graphical illustrations of the error bounds that Willig (1976) sought to define in discussing the limits of conventional measures of consumer surplus. (For further discussion of these concepts see Freeman 1979 and Just, Hueth, and Schmitz 1982.) For our purposes it is important to highlight a few implications of the methods used to estimate these benefit measures and their direct implications for the practices of BCA.

In many cases, the commodities affected by the action are not traded on organized markets. Indeed, one motivation for public-sector involvement is their lack of availability through private means. Consequently, an important dimension of benefit-cost analysis is finding some method for measuring the benefits associated with changing the availability of these nonmarketed commodities. This issue is especially important because it has been the source of considerable confusion among critics of BCA. The absence of organized markets for goods or services does not in itself imply that consumers do not value them. Moreover, it also does not imply that economists' methods for measuring these values are arbitrary or that the estimates are the economists' personal valuations. This may seem obvious, but Baram (1980) and other critics of BCA could be interpreted as basing their arguments on such assumptions.

A more appropriate description is that benefit analysis often resembles detective work (National Academy of Sciences 1975). The analyst must appraise whether the actions of individuals (or firms) in their observable resource allocation decisions provide clues about their respective valuations of the nonmarketed goods or services. As with any form of detective work, there are several ways to find these clues. One means of classifying the approaches used to estimate these valuations is according to the assumptions they require and the degree to which they recognize the behavioral adjustments which households and firms can take in response to changes in the levels of these nonmarketed goods. Each of the methods to estimate values requires that there be some basis for linking the nonmarketed commodity to a marketed good or an action undertaken by households that can be observed. Of course, observation in this case usually means that the action involves market transactions. Linkages can be technical (or physical) or they can be behavioral. For the former we assume any change in the level of the nonmarketed commodity leads to a change in some other good or service. For example, improvements or expansions of a highway reduce travel times to work. Reductions in air pollution may reduce detrimental health effects or increase the yields of agricultural commodities. The relationships are physical or technical in nature. By contrast, the other potential linkage is behavioral; it maintains that individuals are aware of the nonmarketed commodity and make decisions on one or more marketed goods because of it. This approach assumes the motivations for the jointness in these decisions can be either because of individual preferences or because of the way the nonmarketed and marketed commodities are "delivered" to each person. In the first case, an individual may always consume the two goods together so that if the marketed good is unavailable or priced in excess of the individual's willingness to pay, there will be no reason for demanding the nonmarketed good. In the second the two may be tied together through supply. The purchase of a home delivers to the buyer not only the physical structure and its land but also the neighborhood and environmental amenities that are associated with its location. To the extent we can model the factors influencing individuals' decisions on the marketed goods it may be possible to infer how the nonmarketed commodities played a role in any observed set of decisions.

The last type of behavioral link sets up a hypothetical institution for valuation; that is, it involves postulating to the individual a hypothetical market and eliciting values for changes in the nonmarketed good as if this market could be constructed. This direct survey or contingent valuation approach maintains that the responses provided under these hypothetical circumstances correspond to what they would be if the actual choices could be made (i.e., the valuations expressed through decisions in actual markets). Figure 2–4 reproduces a figure in Desvousges, Smith, and McGivney (1983) used for classifying these approaches using these con-

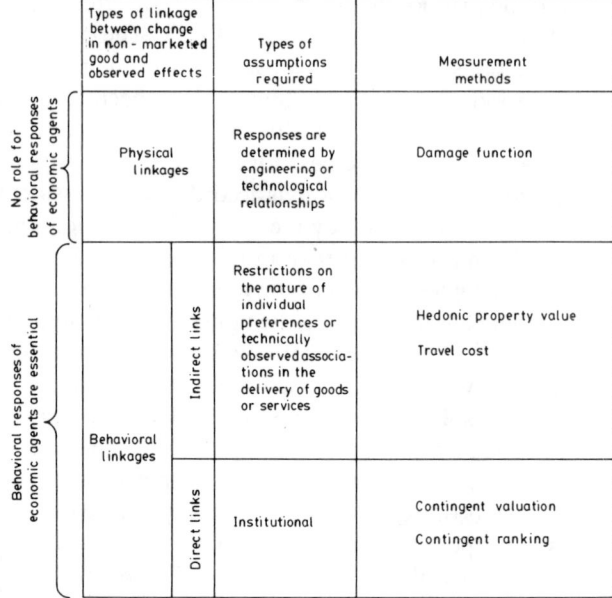

Fig. 2–4. Framework for classifying the methods of measuring economic benefits from nonmarketed goods, from Desvousges, Smith, and McGivney (1983).

siderations. It highlights the key elements in the assumptions made as part of the detective work involved in estimating consumers' values for non-marketed goods and services.

In some cases we do not have methods, as yet, for finding clues as to individual's values of the resources involved. The literature on BCA has generally designated such cases as intangibles (see Haveman and Weisbrod 1975 and Bishop and Cicchetti 1975). This does not imply that they should be omitted from consideration in a BCA. Indeed, a significant component of the research in benefits analysis has been directed toward developing methods for estimating these values. These are the cases where the direct survey or contingent valuation methods for benefit estimation have gained the widest application. It is the absence of a set of circumstances where indirect methods, based on actual behavior, can be used that has in part motivated the development of the contingent valuation approach (see Cummings, Brookshire, and Schulze 1984 for an overview).

Consequently, to criticize the methods of benefit-cost analysis for omitting these intangible effects because they cannot be monetized is incorrect. Their omission may well be the result of an error in a specific benefit-cost study. It is not a fault with the methodology. Moreover, even when estimates are not directly available, there exist a variety of ways of indirectly gauging the importance of the goods and services that are not

2. The Foundations of Benefit-Cost Analysis

valued. Perhaps the most obvious is to pose the question originally raised by Fisher, Krutilla, and Cicchetti (1972) in discussing the preservation benefits of the Hells Canyon reach of the Snake River – namely, how great would the benefits foregone *need to be* to make society indifferent to whether it undertakes the action? There are often indirect or crude indices that can be used to evaluate the plausibility of this excess of costs (associated with losing the preserved environment or whatever is incapable of being measured) over measurable benefits (in the Hells Canyon case, development), thereby answering this question without actually measuring the value of intangibles. Equally important, and often overlooked, is the fact that when measures of the value of intangibles would serve only to reinforce the decision derived by positive net benefits, then they are unnecessary for a "yes-no" judgment as to the potential improvement associated with a project.[5]

Finally, the initial focus on price changes also obscured an important aspect of measuring aggregate benefits. The aggregate benefits we assign to any project providing goods or services outside a market depends on how access to these services is rationed. If we assume all have equal access at fixed (or zero) price, and no binding capacity constraints, then our welfare measures as described earlier are appropriate. However, when access is rationed by some other mechanism, those with maximum willingness to pay may not be assured access and consequently aggregate benefits should not be measured as if they were (see Mumy and Hanke 1975 and Hanke 1981). This point is especially important for cases where congestion or capacity constraints are expected to be important in the provision of the services associated with the action.[6]

The Concept of Costs

Ideally, costs are to be measured by the *opportunity costs* of the resources used in the allocation decision. While economists have tended to regard the cost estimation component of a BCA as the easiest, this may be a legacy of the early project orientation of the analysis. That is, when the action involves building a dam, improving a recreational facility, or modifying a stream to reduce the prospects for flooding, engineering estimates of the costs of the project are constructed as part of its design. While there may be technical issues associated with the treatment of capital and operating costs, these tasks are more direct than many benefit estimation problems. However, an increasing share of BCA activity has been directed to actions, such as regulations, that do not involve direct government provision of goods or services but rather indirect controls on how the private sector provides goods or services. Most health, safety, and environmental regulations would fall in this category. In this case, cost estimation requires second-guessing how the private sector will comply with each regulation. Even in the case of the so-called technology-based regulations, costs will vary with the type of plant, size, etc.[7] Information on these costs may be quite difficult to acquire, since firms have incentives to alter their responses or to conceal actual operating information.

Two methods for cost estimation have been considered in BCA: econometric cost models and engineering estimates. The majority of studies, both in specific project analyses and in evaluation of regulations, have relied on the latter. As a rule, the cost estimates needed are either too specific or detailed to be consistent with the more general ones which could be developed from econometric models, largely because of the state of the art of neoclassical modeling and data limitations (see Kopp and Portney 1983 for further discussion).

An important set of difficulties with the engineering estimates associated with BCA applied to regulatory policies arises from the methods used to construct them. They are routinely based on a model plant format. That is, a first step in the estimation process requires a specification of the features of the plant as they exist before the action is taken. For example, in the case of cost estimates associated with air quality regulations, steam generating electric plants might be classified by the size of the generating capacity (in megawatts), the fuel source (i.e., coal, oil, or gas, often with specific details on the sulfur and ash content of the coal used), and the heat rates. Judgments specifying comparable plants, and often product mix descriptions, are used for other industries. This approach is more consistent with *ex post* cost estimation. That is, it describes how the firm or plant would respond, with existing capital equipment, to the regulation. While it has been used, the process is much more difficult in the case of the costs associated with new plants – the *ex ante* costs. For *ex ante* cost estimation, the response to the regulatory action under evaluation could be incorporated as part of the design of the plant and, as a result, would affect all of the capital selected for it.[8]

What is important to the BCA is the increment to costs associated with the action under study. This implies that the cost estimates must in both cases (*ex ante* and *ex post*) specify a baseline or reference position. To the extent the evaluation takes place over time, cost estimation will require both *ex ante* and *ex post* estimates as existing plants are scrapped and replaced. Thus, the baseline is not simply a single definition of a model plant (or set of plants) but can involve a time sequence of plants as they would evolve in the absence of the action as well as a specification of the form they would take with it. Engineering estimates routinely ignore the fact that retirement and scrappage decisions can be influenced by the policies under review, so that the whole nature of what are *ex post* retrofitting versus *ex ante* plant design costs can be affected by the action.

Costs are routinely classified into capital, and operation and maintenance costs. The capital costs are associated with the capital investments required to retrofit an existing plant or the increments in design capital costs for a new plant. Replacement investments and scrappage values are generally treated as exogenous to the action under evaluation.

There have been few evaluations of the quality of the engineering cost estimates in specific benefit-cost applications. What is available suggests that the cost estimates can be exceptionally sensitive to the modeling assumptions used to define them. Changes in the detail of the description of processing options can lead to quite substantial differences in estimates of costs; see Smith and Vaughan (1980) as an example.

For the most part, this discussion has focused on the estimation of costs at a micro level – for the individual plant or firm. BCA requires estimates of the aggregate incremental costs. This often implies more than simply adding up the increments. Some plants cannot or will choose not to comply with the action. There will be plant closures. While the costs (i.e., unemployment effects on localities) of these closures are not included as specific numerical costs, this does not imply that the BCA methodology ignores them. Benefit-cost analysis in the ideal assumes a frictionless world, in which displaced resources from one activity will be reallocated to other activities. In the real world this is not the case. Therefore, an important component of BCA is an evaluation of the extent of these dislocation and adjustment costs – that is, what is the "distribution" of effects by region and sector? Often a component of what is designated an economic impact analysis, these estimates are also a legitimate component of a full BCA, because they provide the basis for judging the assumptions which underlie the use of the BCA and one component of the set of effects implicitly included in the Kaldor-Hicks compensation assumption.

Nonetheless, it should be recognized that the quality of existing cost estimation practices is not an issue that relates exclusively to benefit-cost analysis. It is also directly relevant to the interpretation and argued superiority of cost effectiveness as a basis for evaluating policy actions. That is, to the extent that the rationale for relying on cost effectiveness over BCA relies on the high quality of cost estimates and lower quality of benefit estimates, this conclusion is not warranted for most cases involving the evaluation of regulations. Indeed, the more relevant issue is how to reflect the uncertainty in both cost and benefit estimates so as to enhance their value for policy purposes.

The Role of Time and Uncertainty

Up to this point the description of cost-benefit analysis has largely been abstracted from the fact that an action may lead to benefits and costs that are realized over time. An aggregate net benefits judgment requires that these time-sequenced contributions to net benefits be converted into a common set of units. Even if all the benefits and costs can be described in monetary terms, a dollar today is not the equivalent of a dollar next month or next year. BCA has generally accepted the use of a present-value criterion for temporal aggregation. Two aspects of this simplication should be identified. First, BCA generally assumes that preferences do not change over time. Future generations' preferences for a clean environment, flood protection, or energy security are assumed to be the same as today's. Of course, as a practical matter, many (and certainly most of the early) BCAs involve time horizons that do not extend beyond one generation. Thus, this issue concerns whether we assume an individual's preferences change over his lifetime. It may be more palatable to assume they do not in this latter case.

Of course, invariant preferences do not imply that real willingness to pay for a good or service would be constant. Growth in real income together with positive income elasticities (or elasticities of price flexibility) can be expected to increase willingness to pay. Equally important, systematic changes in costs can result from technological change or increasing scarcity of nonrenewable resources. While these factors can be incorporated through adjustments to the discount rate used to calculate the present value of aggregate net benefits, it is usually easier to model their role in influencing the pattern of change in net benefits over time and to select a discount rate as a separate decision (see Fisher, Krutilla, and Cicchetti 1972).

The second aspect of the process of developing discounted estimates of net benefits relates to the issue of selecting a discount rate itself. Two basic criteria have been proposed as offering the basis for selecting a discount rate: the private opportunity cost of capital and the social rate of discount. Risk, the character of the goods or services provided by the action, the nature of market imperfections associated with the good or closely aligned commodities, and intergenerational "fairness" have been suggested as considerations in the selection of a discount rate.

The discount rate is the central element in present-value conversions of streams of benefits and costs over time into equivalent terms. At the simplest level it has been assumed that *given access to perfect capital markets*, an individual's rate of time preference would be, through the market equilibrium adjustment process, brought into equality with the private opportunity cost of capital (in terms of its marginal productivity) as measured by the foregone opportunities when capital is allocated to one end rather than its next best use. As Arrow (1966:25) observed:

> Thus, the optimal discount rate for public investment is that it uses the rates of return to private capital. But this is only valid in the context of a more inclusive policy in which aggregate capital is being adjusted to a long-run position where the natural rate of interest is governing . . . A perfectly competitive capital market will achieve this optimum.

Lind (1982) used this argument to summarize the first best view of the discount rate we noted earlier. It is not, as he acknowledged, a framework relevant to the real-world selections of a discount rate. In the real world, market imperfections, taxes (such as the corporate income tax) and other

2. The Foundations of Benefit-Cost Analysis

factors can distort the relationship between individuals' time preferences and the market tradeoffs (see Feldstein 1964). Thus, selections must be made in a second-best setting – one that recognizes the need for pragmatic compromise. This compromise must consider:

1. The prospects for differentials in individuals' desired rates of time preferences for private as opposed to public (or collective) goods and services provided from an investment.

2. Differences in the productivity of a dollar invested in the private as opposed to the public sector.

3. The implications of the specific project for the appropriate public-sector attitude toward risk; Arrow and Lind (1970) have argued for risk neutrality under specific conditions that imply a spreading of an essentially constant aggregate risk premium over a large number of individuals. If, however, there is a public-good feature of what is at risk or a correlation between returns to the project under review and the other investments in the economy, this conclusion is not warranted (see Chapter 5).

4. The implications of exceptionally long-lived projects for future generations and intertemporal equity.

A conventional efficiency argument would seem to imply exclusive reliance on the private opportunity cost of capital. That is, if public investment can be assumed to displace private alternatives, then the rate of return to investments of comparable risk and maturity in the private sector would be the relevant opportunity cost of public-sector investments. This conclusion maintains that there are comparable activities taking place in the private sector and that there are no other public goods provided by the action which may have special features that warrant differential treatment. By contrast, Lind's recent excellent summary of this literature (1982) argues for the use of a social rate of discount which reflects the rate at which society is willing to exchange present consumption for future consumption. Moreover, his proposal does not ignore the question of public investments displacing private. Rather, following Bradford (1975), he calls for reflecting these effects in measurement of the opportunity cost of capital through a shadow price of capital. This shadow price considers the discounted present value (at the social rate) of dollars allocated to the public investments. Since some of these might have been consumption expenditures, this is not simply a stream of private investment returns.

Of course, as we indicated above, the selection of a discount rate cannot ignore the extent and type of uncertainty in the outputs of the action. In effect, one must ask whether, given that all the individuals in a community are risk averse and that a public action will yield an uncertain flow of services, what attitude toward risk should be used in its evaluation by the public sector? In this case, there are also arguments for ignoring risk and adopting a risk-neutral rate even for circumstances in which the private sector would accept a risk-averse rate. Lind (1982) has integrated those various positions by noting that the selection of a discount rate, and with it an attitude toward risk, depends upon how the action under review affects (and if affected by) other public and private investments. Actions that seek to diversify the implied portfolio of investments should be less that the private rate. Otherwise, risk alone cannot justify departure from private opportunity-cost criteria.

As a practical matter, each action will have attributes that prevent uniform recommendations for discount rate. What can be said is that the analysis should be conducted with several rates, so that the sensitivity of the BCA decision to the discount rate selected can be determined. Unfortunately, this does not tell the analyst where to begin in defining this range. Alternative approaches to discounting are discussed in Chapter 5.

Lind (1982) concluded his overall review of the discount rate literature noting that based on after-tax rates of return, an approximate estimate of the risk-adjusted, real rate of return for projects with the same risk as the market portfolio would be 4.6 percent. Moreover, for energy-related investments (his specific focus), considering their relationship to the market, he argued that the appropriate range for the real rate of discount, *after adjustment for their respective impacts on private investment* (using the shadow price of capital; see Arrow 1966 and Bradford 1975), was between 2 and 4.6 percent. While substantially below most of the rates used in current analyses, this range with Lind's specified adjustments to other components for the analysis is likely to find substantial support among a broad spectrum of economists and provides a reasonable starting point for a wide range of public investment decisions that can be argued to have features similar to those considered by Lind. It is not clear, however, how applicable this analysis is to health and safety investments.

Lest this description thus far seem routine, it is important to recognize that in an agency context the responsibility for the individual components of a benefit-cost analysis is often assigned to different units within the agency. This implies that some of the implicit assumptions of BCA may not be upheld for some policy applications. Equally important, it implies that there can exist the prospects for inconsistencies in the estimates of benefits and costs that introduce additional sources of error into attempts to integrate each set of estimates to derive the aggregate net benefits of the action. Thus, the design of strategies for preparing benefit-cost analyses must recognize the role of institutional factors as well as economic principles for the processes determining the quality of the information BCA can provide.

Benefit-Cost Analysis in Comparison with Other Frameworks

Policy decisions are made in response to a number of criteria. They are generally guided by the enabling legislation for the department or agency defining the policy or by legislation affecting specific tasks assigned to them. As Baram (1980) has noted, such legislation, especially for social regulations, has had a tendency to set broad goals and to leave the details of implementation to the administrative rule-making process.[9] Since most legislation is the result of compromise, it will tend to have multiple objectives and can include contradictory goals. The rule-making process must resolve contradictions and define a decision process together with a set of information requirements that responds to the intent of the legislation. Consequently, what have been termed decision-making frameworks are best treated as means of organizing information to display its implications under a preassigned set of criteria. None of the frameworks should be viewed as providing the equivalent of an automatic basis for decisions or as supplanting the decision *process*. (See Chapters 3–5 for further discussion of the process perspective.) Benefit-cost analysis (as it has been defined in this chapter) organizes information in terms that display the economic efficiency implications of an action. Other approaches emphasize one or more additional dimensions of the possible goals governing decisions.

Table 2–1 provides a brief description and listing of the advantages and disadvantages of eight decision-making frameworks frequently proposed and/or used in the policy-making process. Their organization is based on whether the information and logic presumed to underlie the framework focuses on a technical display of information or embodies some normative view of how decisions should be made.

The first category, the set of technical criteria, would seem to view the decision process as maintaining a fixed objective (or set of objectives) with little scope for tradeoffs among objectives. *Technology-based standards*, for example, require that an agency select the "best" technology to meet an objective, presumably as if this decision were based simply on a judgment of what constitutes the best engineering practices. In reality, this is not the case. Tradeoffs are made. Judgments of the ability of individual firms, in particular circumstances, to adopt certain modifications in their respective plants or operating conditions must be made. However, these tradeoffs are rarely identified in the information reported on the specified best technical standard (see Ferland 1983).

With *risk-risk analysis* again the objective seems to be reducing or at least holding constant the levels of risk imposed on individuals. Judgment enters in this case in the evaluation of what are comparable types of risks. Finally, for *cost effectiveness analysis*, a predefined set of objectives is specified and approaches evaluated according to their respective costs of meeting the objectives. No consideration is given to trading off one objective for another in the selection of the least cost approach.

By contrast, the normative approaches all incorporate or lend themselves more readily to incorporating a method for trading off among objectives as a

Table 2–1. An overview of decision-making frameworks*

Conceptual basis/method	Description	Advantages	Disadvantages	Additional references
A. *Technical*				
1. Technology-based standards	Defines the "best" usually in engineering terms; basis for realizing specific objective	Limited informational requirements; requires technical engineering information consistent with informed judgment of best industry practices	"Best" is never clear-cut, even in engineering context; must appraise ability of firms in industry to adopt in order to judge best industry practice; usually undertaken for limited number of design cases (e.g., model plants)	Freeman (1978), Ferland (1983), Grubb, Whittington, and Humphries (1984)
2. Risk-risk analysis	Evaluates policy based on change in risks under status quo versus policy; does new policy introduce, reduce, or substitute risks for existing practices	Forces comparison of policy outcome with existing circumstances; reconizes trade-off in risk-related policies; limits informational needs to those associated with the required risk assessments	Actions can have effects on variables other than simply risk; risks that are compared may not be comensurate; attributes of risk affect individual's reactions to risk changes	NAS (1975), Huber (1983a), (1983b)
3. Cost effectiveness analysis	Selects policy that will minimize costs of realizing the policy goal or objectives	No need to know the benefits; focus is on information often more readily available; provides implicit values of objective (e.g., marginal cost of increasing by one unit)	No consideration given to relative importance of outputs; degree to which *all* costs considered will be important to judgments as to "best" approach; how to treat social costs resulting from side effects	Pearce and Nash (1981)

2. The Foundations of Benefit-Cost Analysis

Table 2–1 (continued)

Conceptual basis/method	Description	Advantages	Disadvantages	Additional references
B. *Normative*				
1. Benefit-cost analysis	Evaluates policies based on a quantification of net benefits (benefits-costs) associated with them	Considers the value (in terms of what individuals will pay) and costs of actions; translates outcomes into commensurate terms; consistent with judging action by efficiency implications	No direct consideration of distribution of benefits and costs; significant informational requirements; tends to omit outputs whose effects cannot be quantified; tends to lead to maintenance of status quo; contingent on existing distribution of income and wealth	NAS (1975), Mishan (1976), Pearce and Nash (1981), Desvousges and Smith (1983), (1984b)
2. Risk-benefit analysis	Evaluates the benefits associated with a policy in comparison with its risks	Framework is left vague for flexibility; intended to permit consideration of all risks, benefits and costs; not an automatic decision rule	Too vague; factors to be considered commensurate are not	Starr (1969); Fischhoff *et al.* (1981)
3. Environmental impact statement	Detailed statements prepared as part of the National Environmental Policy Act of an action's impacts, adverse effects, alternatives; requires a balancing of economic and environmental benefits and costs	Explicitly requires consideration of environmental effects; ability to monetize does not preempt enumeration of all aspects of benefits and costs of an action	Difficult to integrate descriptive analyses of intangible effects with monetary benefits and costs; no clear criteria for using information in decision	Andrews (1982), (1984)

Table 2–1 (continued)

Conceptual basis/method	Description	Advantages	Disadvantages	Additional references
4. Economic impact analysis	Evaluation of the effects of an action on prices, outputs, employment, and other economic features of industries, regions, and governmental units	Many methods for evaluating actions, ignores adjustment costs; provides basis for gauging nature of these costs and what groups (e.g., industries, regions) will gain and lose from action	Economic methods limited because based on equilibrium concepts, and adjustment is disequilibrium phenomemnon; no clear criteria for using information in decision	Lovell, Knox, and Smith (1985)
5. Multiple objective programming	Uses of mathematical programming techniques to select projects based on objective functions including weighted goals of decision maker, with explicit consideration of constraints to action and costs	Offers consistent basis for making all project or regulatory decisions; fully reflects goals and constraints incorporated in model; allows quantification of implicit costs of constraints; permits prioritizing of projects	Results only as good as inputs to model; unrealistic characterization of decision process; must supply the weights to be assigned to goals; large information needs for quantification	Cohon (1978), Zeleny (1982)

* All of these methods, with the exception of environmental impact statements and multiple-objective programming, are described in Chapter 2 of Lave (1981), which provides a brief, nontechnical description of the other methods in this table along with other decision criteria.

part of the analysis of actions. The most direct examples of this are *benefit-cost analysis* and *multiple objective programming*. In other cases, the basis for tradeoffs is vague. *Economic impact analysis* as an instrument for implementing the requirements of the Regulatory Flexibility Act would be an example where the basis of these tradeoffs is least well defined.

These techniques do not represent mutually exclusive alternatives. Indeed, our earlier discussion of the definition of costs for a benefit-cost analysis suggested that economic impact analyses could provide the basis for judging the importance of some of the assumptions underlying BCA. However, this does not imply that the results of several different frameworks, each undertaken independently of the others, can be provided as summaries of the information relevant to specific objectives that are to guide policy making. The objective of providing an integrated appraisal of an action displaying information relevant to several goals must be established in the design of the policy analysis process. It cannot be an afterthought. Without this design consistency, the assumptions of frameworks can be incompatible. Moreover, inconsistencies in the assumptions used for each framework are not necessarily or exclusively the result of defining the problem differently but are often the result of the process of translating the action under evaluation into terms that can be accommodated by the available methods (and empirical information) of each decision framework. Indeed, this is often cited as one advantage of multiple objective programming. Since it is a specific mathematical formulation of the decision process, it requires consistency in the summary of how actions affect each goal.

Implications

Our description of benefit-cost analysis has necessarily been general and as such cannot account for the specific details that affect the quality and interpretation of benefit-cost results in each application. A recent evaluation of the use of benefit-cost analysis under Executive Order 12291 for environmental regulations (Smith 1984c) suggests that the requirements imposed on BCA can exceed our current understanding of existing BCA practices and may well require substantial professional judgments in areas of analysis where there may be no professional consensus on the appropriate method for dealing with a particular task. Since some specific areas of research have been identified as part of that earlier analysis, they will not be discussed here. Rather, it is important to repeat two overall observations.

The limitations of benefit-cost analysis do not imply the equivalent of a "counsel of paralysis." They identify the importance of (1) quantifying the degree of uncertainty in each set of benefit-cost estimates (in a policy-relevant format), and (2) implementing a program of research that systematically evaluates the consequences of the pragmatic judgments that are part of the benefit-cost analyses required by the Executive Order. A recommendation advocating such evaluations does not imply that the judgments required to perform specific BCAs are inappropriate; rather it is a recognition that there is always scope for improvement. We should recognize the limitations of the tools used to organize information and develop systematic procedures for learning about them based on past experiences in order to improve current practices. This should be a fundamental dimension of any attempt to systematically structure decision-making frameworks for policy purposes.

Vanderbilt University

Notes

More specific details on my work in these areas from which this summary is drawn can be found in benefit-cost assessment practices as they relate to water quality programs, in Desvousges and Smith (1983) and Desvousges, Smith, and McGivney (1983); benefit-cost and risk-assessment comparisons, in Smith (1984a); intrinsic and intangible benefits, in Smith (1984b); and benefit-cost analysis and Executive Order 12291, in Smith (1984c). Overall references on benefit-cost analysis include Mishan (1976), Pearce and Nash (1981), Sugden and Williams (1978), and Dasgupta, Marglin, and Sen (1972).

1. For examples see the National Academy report (1975): 39,164.

2. See Krutilla (1961) for an early discussion of this relationship and Just, Hueth, and Schmitz (1982), especially Chapters 2, 3, and 14, for a detailed discussion of the current literature on the relationship between welfare theory and benefit-cost practices.

3. Of course, we need to also consider the Scitovsky reversal test to assure that the change is a Pareto-superior state; see Just, Hueth, and Schmitz (1982): 37–40.

4. Recent contributions by Hausman (1981), McKenzie (1983), and Vartia (1983) demonstrate methods for deriving, under specific assumptions, estimates of the Hicksian welfare measures with information on ordinary demand functions.

5. This would not be the case if BCA methods were used in the calculation of benefit-cost ratios or internal rates of return. This chapter has deliberately avoided consideration of ranking alternative actions on the basis of such summaries, because this was not the original intention of BCA.

6. See McConnell and Sutinen (1984) for a discussion of these issues in relationship to benefit estimation with recreation demand models.

7. See, e.g., the discussion of the cost estimates associated with the best-available-technology standards for water-based residuals with the iron and steel industries in Grubb, Whittington, and Humphries (1984) and Ferland (1983).

8. See Vaughan, Russell, and Cochrane (1976) for a comparison of the estimates of the costs of compliance to environmental regulations under these two cases using a large-scale process analysis model for an iron and steel plant.

9. For an interesting discussion of the motivations for this approach to the design of such legislation, see Zeckhauser (1981).

References

Andrews, Richard N. L. 1982. 'Cost Benefit Analysis as Regulatory Reform.' In *Cost Benefit Analysis and Environmental Regulations*, D. Swartzman, R. A. Liroff, and K. G. Croke, eds. Washington, D.C.: The Conservation Foundation.

———. 1984. 'Economics and Environmental Decisions, Past and Present.' In *Environmental Policy Under Reagan's Executive Order: The Role of Benefit-Cost Analysis*. V. Kerry Smith, ed. Chapel Hill: University of North Carolina Press.

Arrow, Kenneth J. 1966. 'Discounting and Public Investment Criteria.' In *Water Research*, A. V. Kneese and S. C. Smith, eds. Baltimore: Johns Hopkins University Press.

——— and Robert C. Lind. 1970. 'Uncertainty and the Evaluation of Public Investment Decisions.' *American Economic Review* 60: 364–378.

Baram, Michael S. 1980. 'Cost-Benefit Analysis: An Inadequate Basis for Health, Safety, and Environmental Regulatory Decisionmaking.' *Ecology Law Quarterly* 8: 473–531.

Bishop, John and Charles J. Cicchetti. 1975. 'Some Institutional and Conceptual Thoughts on the Measurement of Indirect and Intangible Benefits and Costs.' In *Cost Benefit Analysis and*

Water Pollution Policy, Henry M. Peskin and Eugene P. Seskin, eds. Washington, D.C.: The Urban Institute.

Bradford, David F. 1970. 'Benefit-Cost Analysis and Demand Curves for Public Goods.' *Kyklos* 23: 1145–1159.

———. 1975. 'Constraints on Government Investment Opportunities and the Choice of Discount Rate.' *American Economic Review* 65: 887–895.

Cohon, Jahred L. 1978. *Multi-Objective Programming and Planning*. New York: Academic Press.

Crouch, E. A. C. and R. Wilson. 1982. *Risk/Benefit Analysis*. Cambridge: Ballinger.

Cummings, Ronald G., David S. Brookshire, and William D. Schulze. 1984. 'Valuing Environmental Goods: A State of the Arts Assessment of the Contingent Valuation Method.' Unpublished draft monograph. University of New Mexico.

Dasgupta, Partha, Stephen Marglin, and A. K. Sen. 1972. *Guidelines for Project Evaluation*. New York: United Nations.

Desvousges, William H. and V. Kerry Smith. 1983. *Benefit Cost Assessment Handbook for Water Programs*, Vol. I. Prepared for Environmental Protection Agency. Research Triangle Institute.

———, ———, and Matthew McGivney. 1983. *A Comparison of Alternative Approaches for Estimating Recreation and Related Benefits of Water Quality Improvement*. Environmental Benefits Analysis Series. Washington, D.C.: Environmental Protection Agency.

Feldstein, Martin S. 1964. 'The Social Time Preference Discount Rate in Cost-Benefit Analysis.' *Economic Journal* 74: 360–379.

Ferland, Kathey A. 1983. 'Benefit Cost Analysis in Environmental Decision Making: A Case Study of the Implementation of EO 12291 in the Environmental Protection Agency.' Masters thesis. University of North Carolina, Chapel Hill.

Fischhoff, Baruch, Sarah Lichtenstein, Paul Slovic, Stephen L., Derby, and Ralp h L. Keeney. 1981. *Acceptable Risk*. Cambridge: Cambridge University Press.

Fisher, Anthony C., John V. Krutilla, and Charles J. Cicchetti. 1972. 'The Economics of Environmental Preservation: A Theoretical and Empirical Analysis.' *American Economic Review* 62: 605–619.

——— and V. Kerry Smith. 1982. 'Economic Evaluation of Energy's Environmental Costs with Special Reference to Air Pollution.' *Annual Review of Energy* 7: 1–35.

Freeman, A. Myrick III. 1978. 'Air and Water Pollution Policy.' In *Current Issues in U.S. Environment Policy*, Paul R. Portney, ed. Baltimore: Johns Hopkins University Press.

———. 1979. *The Benefits of Environmental Improvement: Theory and Practice*. Baltimore: Johns Hopkins University Press.

Grubb, W. Norton, Dale Whittington, and Michael Humphries. 1984. 'The Ambiguities of Benefit Cost Analysis: An Evaluation of Regulatory Impact Analysis Under Executive Order 12291.' In *Environmental Policy Under Reagan's Executive Order; The Role of Benefit-Cost Analysis*, V. Kerry Smith, ed. Chapel Hill: University of North Carolina Press.

Hanke, Steve H. 1981. 'On the Feasibility of Benefit-Cost Analysis.' *Public Policy* 29: 147–57.

Hausman, Jerry. 1981. 'Exact Consumer's Surplus and Deadweight Loss.' *American Economic Review* 71 (September): 662–676.

Haveman, Robert H. and Burton A. Weisbrod. 1975. 'The Concept of Benefits in Cost-Benefit Analysis; With Emphasis on Water Pollution Control Activities.' In *Cost Benefit Analysis and Water Pollution Policy*, Henry M. Peskin and Eugene P. Seskin, eds. Washington, D.C.: The Urban Institute.

Huber, Peter. 1983a 'The Old-New Division in Risk Regulation.' *Virginia Law Review* 69: 1025–1107.

———. 1983b. 'Exorcists versus Gatekeepers in Risk Regulation.' *Regulation* (November. December): 23–32.

Just, Richard E., Darrel L. Hueth, and Andrew Schmitz. 1982. *Applied Welfare Economics and Public Policy*. Englewood Cliffs, N.J.: Prentice-Hall.

Kopp, Raymond J. and Paul R. Portney. 1983. 'Estimating Environmental Compliance Costs for Industry: Engineering and Economic Approaches.' In *Proceedings of Workshop on Effects of Environmental Regulation on Industrial Compliance Costs and Technology Innovations*. Washington, D.C.: National Science Foundation.

Krutilla, John V. 1961. 'Welfare Aspects of Benefit-Cost Analysis.' *Journal of Political Economy* 69: 226–235.

Lave, Lester B. 1981, *The Strategy of Social Regulation*. Washington, D.C.: Brookings Institution.

Lind, Robert C. 1982. 'A Primer on the Major Issues Relating to the Discount Rate for Evaluating National Energy Options.' In *Discounting for Time and Risk in Energy Policy*, Robert C. Lind, ed. Baltimore: Johns Hopkins University Press.

Lovell, C., A. Knox and V. Kerry Smith. 1985. 'Microeconomic Analysis.' In *Climate Impact Assessment: Studies of the Interaction of Climate and Society*, Robert Kates, ed. New York: Wiley.

McConnell, Kenneth E. and Jon Sutinen. 1984. 'A Conceptual Analysis of Congested Recreation Sites.' In *Advances in Applied Micro-Economics*, Vol. III, V. Kerry Smith and Ann D. Witte, eds. Greenwich, Conn.: JAI Press.

McKenzie, George W. 1983. *Measuring Economic Welfare: New Methods*. Cambridge: Cambridge University Press.

Mishan, E. J. 1976. *Cost-Benefit Analysis*, rev. ed. New York: Praeger.

Mumy, G. E. and S. H. Hanke. 1975. 'Public Investment Criteria for Underpriced Public Products.' *American Economic Review* 65(4): 713–720.

National Academy of Sciences. 1975. *Decisionmaking for Regulating Chemicals in the Environment*. Washington, D.C.: National Research Council.

Pearce, David W. and C. A. Nash. 1981. *The Social Appraisal of Projects: A Text in Cost-Benefit Analysis*. New York: Wiley.

Smith, V. Kerry. 1984a. 'Environmental Policy Making under Executive Order 12291: An Introduction.' In *Environmental Policy under Reagan's Executive Order: The Role of Benefit-Cost Analysis*, V. Kerry Smith, ed. Chapel Hill: University of North Carolina Press.

———. 1984b. 'Intrinsic Values in Benefit Cost Analysis: New Research Frontier or License for Abuse.' Working Paper No. 84–W19. Vanderbilt University.

———. 1984c. 'The Integration of Benefit Cost Analysis and Risk Assessment.' Working Paper No. 84–W15. Vanderbilt University.

——— and William J. Vaughan. 1980. 'The Implications of Model Complexity for Environmental Management.' *Journal of Environmental Economics and Management* 7: 184–208.

Starr, Chauncey. 1969. 'Social Benefit Versus Technological Risk.' *Science* 165: 1232–1238.

Sugden, Robert and Alan Williams. 1978. *The Principles of Practical Cost-Benefit Analysis*. Oxford: Oxford University Press.

Vartia, Yrjo O. 1983. 'Efficient Methods of Measuring Welfare Change and Compensated Income in Terms of Ordinary Demand Functions.' *Econometrica* 51: 79–98.

Vaughan, William J., Clifford S. Russell, and H. C. Cochrane. 1976. *Government Policies and the Adoption of Innovations in the Integrated Iron and Steel Industry*. Report prepared for the National Science Foundation, National R&D Assessment Program.

Willig, Robert D. 1976. 'Consumer's Surplus Without Apology.' *American Economic Review* 66: 587–597.

Zeckhauser, Richard. 1981. 'Preferred Policies When There Is a Concern for Probability of Adoption.' *Journal of Environmental Economics and Management* 8: 215–237.

——— and Anthony C. Fisher. 1976. 'Averting Behavior and External Diseconomies.' Kennedy School Discussion Paper No. 41D. Harvard University.

Zeleny, Milan. 1982. *Multiple Criteria Decisionmaking*. New York: Wiley.

Chapter 3

Benefits Assessment for Regulatory Problems

Howard Kunreuther and Lisa Bendixen

This chapter is concerned with the role of benefits assesment within the regulatory decision-making framework for public policy problems involving risk. The basis of most government regulation is to minimize risks and detrimental economic impacts. Since risks cannot be exchanged in most instances, it becomes necessary to have a centralized decision process which can regulate the levels of risk. Not all risks can be fully regulated; one must focus on identified and controllable risks and impacts. To motivate the discussion, consider the following four problems of current interest where some type of regulation has been ordered:

1. *Automobile safety*. It is known that seatbelts save lives and that few people wear seatbelts. What are the potential benefits of regulations with respect to reducing injuries and fatalities on the road? Specifically, how would one evaluate the formal requirements that some type of passive restraint (e.g., automatic seatbelt, airbag) be installed on all new cars? This issue currently faces the National Highway Traffic Safety Administration in the Department of Transportation. They must take into account the preferences and behavior of the public, as well as the preferences and costs to the auto industry. Each group would select a different approach to regulation.

2. *Natural disasters*: What are the appropriate roles that regulation should play in reducing potential losses from natural disasters? In particular, are there hazard-mitigation measures that should be required for residents in hazard-prone areas? This problem currently faces the Federal Emergency Management Agency, who must also recognize that individuals often underestimate the probabilities of an accident and therefore are less willing to accept regulations than they would be under more accurate states of knowledge.

3. *Hazardous materials management*: What is the appropriate role of regulations with respect to the transport, storage, and disposal of hazardous materials? How should the government deal with existing and potentially new sites for hazardous waste? What are the costs of prevention compared with the costs of relocation and cleanup after a

release of waste material? These types of problems currently face the Environmental Protection Agency.

4. *Nuclear power plants*. What regulations should be a part of the licensing and operating procedures associated with nuclear power plants? How should the benefits of such regulations be measured given the difficulties in relating exposure to health effects? These problems currently face the U.S. Nuclear Regulatory Commission.

Considerations in Evaluating Benefits of Regulation

In evaluating the benefits of regulations for public policy problems, one must take into account the nature of the problem, the relevant interested parties (in terms of both the overall decision and the benefits), the appropriate decision processes (including the legal basis for decision making), and the available alternatives. Each of these features is discussed below. The interactions of the various features are shown in Figure 3–1.

Problem Formulation

This involves an understanding of the status quo and the performance of current policies utilized to deal with a particular situation. Consider the automobile safety problem. Today no specific regulations are in force regarding the use of seatbelts or airbags, although recent legislation may change this situation in the immediate future. Arnould and Grabowski (1981) indicate that current estimates of voluntary belt usage, based on large-scale field studies, are between 10 and 20 percent, despite the fact that

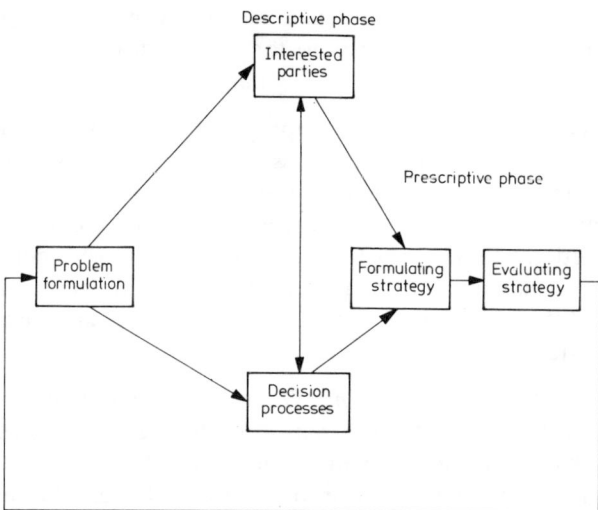

Fig. 3–1. Conceptual framework.

most people are aware that engineering studies have found seatbelts to be highly effective in reducing the probability of death or serious injury from automobile accidents.

For some problems, regulations may be enforced but will only affect a small proportion of the population. For example, the Federal Flood Protection Act requires all home owners who obtain federally insured mortgages to purchase flood insurance if they are in this type of hazard-prone area. One would need to know the proportion of individuals in such areas who are currently insured as a result of this type of regulation.

Interested Parties

In looking at benefits assessments for regulation, one must identify the set of stakeholders who will be affected by a particular regulation either in terms of benefits or impacts on future decisions. Each of the interested parties has its own set of attributes which it considers important with respect to a particular problem area. Furthermore, the weights and importance placed on these different attributes will vary across these parties. For example, with respect to problems of regulating hazardous waste and nuclear power plant siting, the concerns of residents near a proposed facility will be considerably different from those of businesses and industries involved with the problem or communities some distance from the proposed site. Only by delineating the different stakeholders and their concerns can one be in a position to evaluate the potential benefits of regulation in improving or changing the status quo, as well as the equity of the distribution of benefits.

Decision Process

What data are required for evaluating the potential benefits of regulation and how are these data processed? Two elements are particularly important here. First, there is a need to specify the type of statistical data available on the nature of the risk involved. For some events, there is considerable knowledge on both the risk and the role regulations can play in reducing it (e.g., automobile safety). For other risks, particularly those associated with new technological facilities (nuclear power plants, liquefied natural gas storage), there are limited or no statistical data on the probability and consequences of accidents of different magnitudes. The assessment of the expected benefits from different regulatory mechanisms has to take into account the state of knowledge of the risk involved and its degree of uncertainty. The second important element is an understanding of the way this knowledge is processed by different interested parties. There is considerable empirical evidence that for certain problems the public perception of the risk differs from its statistical description. For example, Slovic, Fischhoff, and Lichtenstein (1983) reveal that lay persons perceive many hazards such as home appliances and smoking to be considerably less dangerous than historical data have shown them to be.

In order to determine the potential benefits of a regulation, it is important to know how different individuals behave with respect to a risk. For example, it is well known that individuals fail to protect themselves against events which they perceive to have a low probability of occurrence (e.g., auto accidents, natural disasters). If individuals behave as if the probability is zero – that is, if they employ some type of threshold model – then it is an open question as to whether they will change their behavior if adequate information is presented to them on the true nature of the risk. If not, some type of regulation may be required. Kleindorfer and Kunreuther (1981) postulated that government regulation is most likely to be beneficial and acceptable where public information is inadequate, incorrect, or misunderstood. This might be the case when one is dealing with very complex products, technologies, or hazards which are unfamiliar and where the potential consequences of events with very low probabilities of occurrence are greater than previously realized by the public or by industry and government. Government regulation may also be more acceptable when events which are being regulated are very unlikely to occur.

Alternatives to Regulations

In evaluating the potential benefits of regulation, there is a need to compare the proposed strategy with other alternatives for the sake of thoroughness, acceptability, and ease of implementation. The Environmental Protection Agency has listed four categories of alternatives which might be considered:

Alternatives to federal regulation. These include negotiated voluntary actions and market, judicial, or state or local regulatory mechanisms that could address the environmental problem.

Alternatives within the legislative provision's scope. These include three broad categories of alternatives: the degree of control, effective compliance dates, and methods of ensuring compliance. The second category, for instance, could involve issuing related regulations jointly to allow the affected industries to modify their products or operations to comply simultaneously with all the regulations. Alternative methods of ensuring compliance could involve employing various enforcement options (e.g., on-site inspections versus periodic reporting and sharing implementation responsibilities among the different levels of government); using various compliance methods for different segments of industry or types of economic activity where costs of compliance vary sharply (e.g., different treatment for small and large firms); or allowing variances under certain conditions.

Market-oriented regulatory alternatives (whether or not they are explicitly authorized in the Agency's legislative mandate). These methods include using information or labeling to enable consumers or workers to evaluate hazards themselves and using economic incentives, such as fees or charges, marketable permits or offsets, changes in insurance provisions, or changes in property rights.

Major alternatives beyond the scope of the legislative provision under which the proposed regulation is being promulgated. These would include regulatory alternatives that controlled other routes of exposure and that possibly would be authorized under other legislation. [EPA 1983: 5.]

In the context of one of the examples presented at the outset of this chapter, the automative industry could offer a subsidy to the family of anyone killed while wearing a seatbelt (as is currently being done for buyers of new General Motors cars), instead of providing automatic restraint systems through regulation. The proposed national automobile safety regulation has provisions in it which encourage alternate state-level regulations. These two approaches represent, respectively, market-oriented regulatory alternatives and alternatives to federal regulation.

Benefits assessment does not give a final ordering of alternatives but does give a ranking along one of the characteristics which may contribute to the final ordering and selection of alternatives. When alternative strategies are being evaluated, there are a number of objectives. These include the overall desire to improve the well-being of society and the desire to have equitable distributions of risks and benefits. Often these may be conflicting objectives. In addition, the feasibility and efficiency of any given strategy should be considered, something that is rarely done within the context of a benefits assessment.

The next section provides a rationale for looking at the benefits of regulation in the context of Executive Order 12291. Then a five-step process is developed for structuring a benefits assessment for particular regulatory problems, and detailed steps are presented for conducting a benefits assessment.

Benefits Assessment within the Policy Process

The role that the government and its regulatory agencies is supposed to play varies according to whom one asks. Some feel that the government is bound to make the same decision an individual would if he had the power and responsibility to do so, while others feel that any worthwhile public decision must be based on ethical judgments above and beyond consumer sovereignty (Jones-Lee 1976). The more uncertain the scientific basis of the data, the greater the need will be for flexibility and adaptability in the analysis and the decision-making process. More discretion will also have to be left to the regulatory agency in choosing the preferred course of action (Majone 1983). There may even be occasions where certain governmentwide decisions are justified that totally supersede the need for any type of detailed analysis.

The need to regulate arises for many reasons, including the fact that many actions affect or potentially affect individuals in ways which they are either unaware of or unable to prevent. In order to take corrective action from exposure to risk, an individual must be able to identify the threat, to obtain and process information necessary to make an appropriate decision about it, and be able to remove himself and his family from a contaminated environment if necessary. When these steps are beyond the capabilities or financial resources of an individual, this points to the government as the only reasonable institution to protect the public from harm that they cannot protect themselves from (see Kleindorfer 1982). In addition, a particular decision will often affect groups whose goals or objectives do not coincide. A framework is then required in which these different goals as well as different value judgments can be handled. Cost-benefit analysis can be set up in such a way that varying attitudes toward uncertainty and value tradeoffs between conflicting objectives can be handled (Lathrop 1982), but it cannot resolve the underlying conflict.

There is a need for such a framework today, since regulatory agencies have been criticized for holding back technical progress and negatively influencing the economy, particularly small businesses. Critics contend that there is a perception that certain interests are subject to favoritism, agencies are ineffective, and undue costs are imposed (Lave 1981). As stated by W. Nordhaus, the "roots for reform of the regulatory process are essentially a marriage of better process and better analysis" (Ferguson and LeVeen 1981: 134).

The application of benefits assessment provides inputs to a decision, but assumptions and judgments are still required and the final decision is still a political one. However, the application of benefits assessment can help to achieve regulatory openness by allowing the information available from both government and industry to be reviewed publicly. Problems which make decisions especially sensitive to political judgments include: extreme difficulties in evaluating "intangibles" such as environmental protection and life-saving programs; high consequence – low probability events; inequitable distributions of costs or risks; large-scale impacts; novel technologies or applications (Lathrop 1982). The role of benefits assessment as a provider of "scientific" evidence in a political decision opens it up to the potential for misuse or misinterpretation. This potential is felt to be minimized if the analysis is formalized as a part of the political decision-making process (Fischoff 1977). However, care must be taken to maintain a just process.

There are also questions of government paternalism. Is it fair to force someone who already protects himself by wearing a seatbelt to in addition buy a car with an airbag? Can one fairly impose a regulation on society?

Executive Order 12291 and Judicial Review

Recognizing the improvements offered to regulatory decision making by the inclusion of a rigorous benefits assessment, in 1981 the President issued Executive Order 12291 which states in part (Intro., Secs. 2,3):

> . . . in order to reduce the burdens of existing and future regulations, increase agency accountability for regulatory actions, provide for presidential oversight of the regulatory process, minimize duplication and conflict of regulations, and insure well-reasoned regulations

3. Benefits Assessment for Regulatory Problems

... all agencies, to the extent permitted by law, shall ... in connection with every major rule, prepare, and to the extent permitted by law, consider a Regulatory Impact Analysis.

The general requirements of E.O. 12291 are that decisions be based on adequate information concerning the need for and consequences of a proposed action; that action not be undertaken unless societal benefits outweigh societal costs; that the benefits to society be maximized; that the alternative having the least net societal cost be chosen; that agency priorities be set to maximize net societal benefits considering the status of affected industries, the state of the economy, and future regulations which may be implemented.

A formal procedure for review was also established, as well as a timetable for publication of certain pieces of information. The task force and the Office of Management and Budget were charged with developing uniform standards for conducting Regulatory Impact Analyses and for identifying "duplicative, overlapping and conflicting rules" among other things.

The views within the different agencies of the impact of this ruling vary but are now posed positively. There is a feeling that internal oversight of the entire decision-making process will improve (Snyder 1984). The Nuclear Regulatory Commission (1984) feels that analyses can help in determining whether or not to initiate a regulatory action, providing a coherent, decipherable, and well-documented justification of a particular action, and choosing the best alternative. However, there is still a wide disparity in agency treatments.

Five-Step Procedure in Applying Executive Order 12291

This section develops a five-step procedure building on Lave (1981) for determining how E.O. 12291 can be utililized as a part of the regulatory policy process. As indicated earlier, we feel that the evaluation of the benefits of regulation can only be undertaken by comparing this strategy with alternatives such as market mechanisms and incentive systems. In addition, although these steps may appear to be obvious, it should be remembered that different divisions of an agency may have the responsibility for different steps. The ultimate goal of the procedure is the same as the ultimate goal of E.O. 12291 – to improve decisions.

Step 1: Deciding What to Regulate

This initial step is closely related to problem formulation as described earlier. In addition, there may be legislative mandates, agency histories, and past experience which dictate an evaluation of the benefits and costs of specific regulation. For example, the need for hazardous waste regulation was stimulated by the Love Canal incident; widespread concern with nuclear power regulation resurfaced after the accident at Three Mile Island; the Tylenol scare led to tamper-proof packaging.

Presenting empirical evidence on the passage of safety legislation in the United States, Walker (1977) suggests that accidents and disasters do play an important role in setting the discretionary agenda of political bodies such as the Congress and regulatory agencies. Other examples are provided by Lawless through case histories and problems involving the impact of technology on society. He points out that frequently,

New information of an "alarming" nature is announced and is given a widespread visibility by means of mass communication media. Almost overnight the case can become a subject of discussion and concern to much of the populace and generate strong pressures to evaluate and remedy the problem as rapidly as possible. [Lawless 1977:16.]

While it may be the intent of E.O. 12291 to set priorities on what to regulate based on some analysis of costs and benefits, the selection process is actually often driven by the very political and public attention that Lawless notes. An additional problem is raised when one considers that a regulation regarding one of the areas of safety, health, or the environment can impact on the others; moreover, it can affect the benefits assessment of other regulations within the same area. For example, within the environmental area, when evaluating primary standards for a criteria pollutant how does one treat the New Source Performance Standards for the *same* pollutant?

Step 2: Conducting Analyses

This stage involves detailed analyses regarding potential impacts of regulation on the different interested parties. Figure 3–2 displays the steps involved, which are discussed further in later chapters. It is particularly important to indicate the types of data available and their limitations as well as the opportunity for collecting more information and the costs of such efforts. An important aspect of an analysis is the amount of time available for making a decision. If there are specific deadlines or constraints imposed on the process, the nature of the analysis will be quite different than if it were open-ended. Agencies frequently take a path of compromise in order to minimize conflict, because the regulatory environment is one that puts a premium on answering immediate needs with quick, defendable solutions (Lave 1982).

In all analyses, judgments must be made regarding the suitability of quantification (in dollars or some other units), the treatment of factors which cannot be quantified, and the accuracy of quantified estimates. However, for public policy questions, one is generally more concerned with the soundness of the overall decision-making process than the statistical significance of cost or other estimates.

If the regulation deals with problems that involve potential losses of property or of life, then some type of risk analysis is required. Risk analysis

3. Benefits Assessment for Regulatory Problems

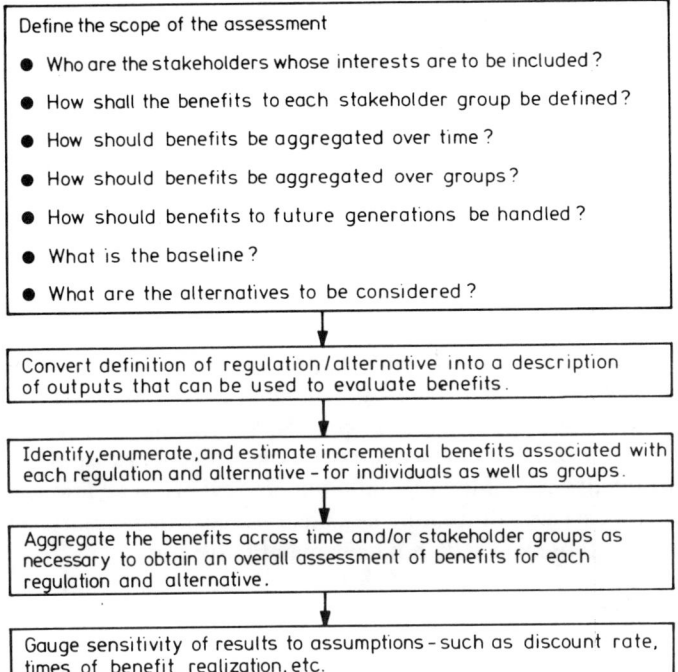

Fig. 3-2. Detailed steps for conducting a benefit-cost analysis.

cannot be viewed in a vacuum but must be examined in the political context of the problem. In recent years the scientific community has lost its impartial image because technological uncertainty has allowed each scientist a wide range of interpretation regarding the risk associated with hazardous technologies. For example, with respect to nuclear power, scientists make judgments regarding the seriousness of health and mental risks that correspond to their attitudes toward nuclear power (Nelkin 1981; Mazur 1982).

While this may be the current result of many studies, various agencies have published fairly thorough guidelines for conducting regulatory impact analyses of high quality and consistency. It is generally recognized that the evaluation of an existing regulation to see if changes are required need not be as detailed as the assessment for a proposed rule-making action, which is in turn less sophisticated than the assessment associated with a full-blown regulatory impact analysis. In general, the level of analytic sophtication should depend on the following, according to the Department of Transportation (1984):

1. The intensity of public and government interest and the associated degree of controversy.

2. The magnitude of individual and societal costs and benefits.
3. The available resources in terms of time, manpower, and budget.
4. The current availability of data, as well as the feasibility and cost of obtaining additional data.

To this list the Nuclear Regulatory Commission adds the complexity of the problem and the closeness of the outcome and stresses that the importance of the action and the availability of data are the most important constraints. A basic principle used in determining the appropriate level of effort is that the resources expended should be commensurate with the value of the information obtained (NRC 1983). In this sense, formalized cost-benefit analysis can actually help in determining an appropriate level of effort.

In any case, it is important to be very explicit about exactly what has and has not been evaluated in the analysis. Benefits assessment cannot measure all second- and greater-order events in almost all cases, and it also does not discuss the cost-effectiveness of the regulation under review. In addition, it does not provide an easy way of treating benefits which can be measured in dollars as opposed to those which do not have quantitative forms.

Step 3: Justifying the Regulation

This step refers to the importance of evaluating the regulation through critical review and public dissemination of appropriate information, a condition that Executive Order 12291 specifically addresses. In conducting this process, it is important to separate out the facts developed in Step 2 from the values of the different interested parties. Currently the public does not believe that its value judgments and objectives are considered or that the analyses are conducted objectively. There is also the feeling that beyond the potential for misuse, benefits assessment goes against ethical and moral standards by trying to place monetary values on life (Zeckhauser and Leonard 1983). Again, this reemphasizes the fact that benefits assessment is just one input into a decision-making process, not a full basis for a decision.

In considering public involvement, there is a need to include groups who are directly affected by some action and a more general group who has an overall interest in a particular area. Although formalizing the role of public involvement and review in the decision-making process can increase public confidence in the eventual decision, it cannot satisfy those who are poorly represented in the process, those whose values are vastly different, or those who distrust the type of assumptions and judgments made (Royal Society 1983). Its effectiveness in the public arena is also dependent on how understandable it is to the general public. A large amount of technical jargon can easily turn off an interested reader.

Additional problems can be introduced as a result of the dynamic and sequential nature of the decision process. Different parties, with varying amounts of power and responsibility, can enter each side of a debate with their own preferences at different stages of the decision process, depending on the issues involved at each stage (Kunreuther *et al.* 1983). This further complicates the identification of affected and interested parties as well as the proper weighting of their values. Each such group may have its own data that it feels should be used and its own constraints that should be met. Benefits analyses can be conducted with a number of different schemes for weighting values, but it cannot "select" the proper set of weights.

For example, a cost-benefit analysis based solely on industry-supplied cost information may be very pessimistic regarding costs. This can push decisions toward maintaining the status quo, thus obstructing the realization of legislative goals (Baram 1980). Some analysts feel that this restricts the use of an economic framework for decision making. Others argue that by looking at the damages due to not regulating, looking at the benefits of particular regulatory incentives taken in the past, and quantifying benefits, those arguments which emphasize costs can be offset (G. Speth, in Ferguson and LeVeen 1981).

Step 4: Evaluating the Regulation

Quantitative analysis can be a useful input to the decision process, replacing vague statements of findings with precisely worded statements of facts and uncertainties regarding the implications of various alternatives. The policy maker, who must select a strategy that reflects the values of society, is confronted with a number of problems in choosing among alternative strategies. First, some measure(s) of welfare must be agreed upon to help rank policies. Second, since such welfare measures can be expected to depend on the actual behavior of consumers and firms, one must also understand what each actor's behavioral response will be to alternative free-market and governmental policies. These problems of misrepresentation of welfare and of predicting behavioral adjustment are central to public policy analysis. At the outset, it is instructive to sketch the traditional framework for dealing with these issues.

In the traditional approach of neoclassical economics, individuals are assumed to be rational. This means that each individual or firm acts as if it were maximizing utility. In the above formulation, all potential victims will be assumed to be identical, with utility function U and initial wealth W_1, the other individuals are also identical and have a utility function V and initial wealth W_2. One of the controversial issues in welfare economics is what weight should be given to different segments of society. In this simplified example, suppose that the policy maker arbitrarily determines that potential victims are given a weight ψ so that other individuals are assigned a weight $1 - \psi$. If each person were to be given equal weight, then $\psi = N_1/(N_1 + N_2)$ and $(1 - \psi) = N_2/(N_1 + N_2)$.

The adjustment process is typically characterized by assuming that potential victims behave as if they were maximizing their expected perceived welfare, taking everyone else's behavior as fixed. The basic data required for the traditional descriptive analysis are the perceptions and associated utility functions of each party. Traditional welfare theory also assumes that each economic agent involved is the best judge of his own utility. There are, of course, many problems in choosing a welfare function, and these issues have been the subject of considerable debate. The economics literature does seem to be in general agreement on one important point, however: the utilities in the welfare function are evaluated under the assumption that individuals and firms are well informed and rational.

Recent work in behavioral decision theory by psychologists and economists suggests that there is considerable misinformation by individuals on risks (see Fischhoff *et al.* 1982) and that people do not behave as if they maximized expected utility (Kahneman and Tversky 1979; Schoemaker 1982). Furthermore, there is a growing body of evidence which suggests that the context in which information is framed will influence preferences (Hershey, Kunreuther, and Schoemaker 1982) and choices (Tversky and Kahneman 1981). One of the reasons for this behavior is that individuals may undertake a form of mental accounting where the same data is perceived differently as a function of how it is presented (Thaler 1980, 1983). These behavioral findings suggest that when evaluating the welfare impacts of alternative programs, it is important to recognize that individuals may not maximize expected utility and that the basic assumptions of economics may be in question.

These findings imply that with respect to each particular problem, it is important to understand how well consumers process information and what type of data they utilize in order to determine whether or not certain regulations improve welfare. Colantoni, Davis, and Swaminuthan (1976) make this point when evaluating policies which impose certain standards (e.g., auto safety features) rather than providing consumers with better information. The alternative methods discussed later in this report may help in situations with insufficient or imperfect information.

Consider the problem of evaluating whether a passive restraint such as airbags should be required as standard equipment for all new automobiles. Kleindorfer and Kunreuther (1981) have analyzed the welfare implications of this proposal in relation to optional installation of this safety feature under different informational assumptions. They show that the citizenry might be willing to make airbags mandatory, if it knew and believed the actual figures, but it might be opposed to such action if consumers underestimate the probability of an accident. Colantoni *et al.* (1976) also point out that any type of regulation may harm a group of consumers who are already taking protective action on their own. For example, those who currently wear seatbelts may be unduly penalized for having to purchase a car with an airbag. This welfare loss has to be compared with a policy of optional standards, whereby many drivers do not avail themselves of protection because they underestimate the risk. A well-structured benefits assessment can improve an individual's information on risks and benefits, but it does not force the individual to change his or her decision process.

The issue of the political feasibility of mandatory requirements has also been analyzed by Pauly, Kunreuther, and Vaupel (1984). Politicians may not support regulations even if they yield potential net benefits to society because the achievement of their goals (which include staying in office) depends on pleasing the same consumers whose misperceptions are the source of the problem. Consequently, compelling the voters to do things which they believe are not in their interest is unlikely to achieve their political support. For this reason, politicians may do nothing with respect to regulations and instead may favor methods such as *ex post* compensation to disaster victims.

This negative attitude toward regulations by politicians may also explain the reluctance of state and Federal agencies to regulate a solution to the hazardous waste facility problem. Rather than forcing a particular community to "accept" a waste treatment plant in its backyard, they prefer efforts to reach settlements through some type of buyer-seller negotiation (see O'Hare, Bacow, and Sanderson 1983; Raiffa 1982). Regulation is most likely to be supported in those areas of public safety, such as the siting of a nuclear power plant, where citizens feel that they have limited understanding of the technical issues and also that there may not be an incentive for developers to make the facility safe enough under traditional market mechanisms including limited liability. Similar arguments could be used for legislation on consumer product safety such as drug regulation, although their potential benefits to consumers have been questioned (Peltzman 1973).

Step 5: Implementation in a Timely Fashion

To really improve social regulation, it is necessary to get regulations implemented promptly without a lot of litigation or unnecessary social costs. When evaluating the overall decision which was reached, the concern is whether the best possible answer was selected given the time, resources, and information available. In the area of public policy, timeliness can be much more important than completeness or accuracy (Lave 1981). As such, the original choice of regulation (Step 1) must consider the ease of implementation of the regulation.

The amount of time available can affect the completeness of scientific evidence. It is quite possible that an analysis to collect and process additional data would be started and that its results would not be available until after a decision had already been made. Furthermore, even if the evidence finally supports another alternative, it can be very difficult to change an action once it has been agreed upon and presented to the public. Therefore, a timely decision must often be made without adequate data, even though mistakes made in the initial stages can be costly and agencies must convince the public of the soundness of their decisions. The output of a benefit assessment should point out any assumptions, alternate interpretations, and data gaps that could not be resolved in the available time frame. Moreover, the methodology used to measure benefits must be selected based on the time, resources, and data available.

Conclusion

This chapter has addressed the importance of understanding the descriptive analysis of a particular problem as a prelude to undertaking benefits assessment of regulation. At the same time, there is a need to understand the decision processes of the relevant stakeholders with respect to a particular problem under consideration. It is generally felt that a carefully applied benefits assessment can cause decisions to be made which do not necessarily maintain the status quo or reflect the desires of the loudest and most powerful stakeholders. The fact that not all concerns are always quantifiable is not a problem unless it produces a bias in the results, because the analysis itself does not produce the final decision but instead is merely a piece of evidence or a fact to use in making what is ultimately a political decision. The next chapter develops a framework which builds on the concepts of this chapter and the previous one.

University of Pennsylvania
Arthur D. Little, Inc.

References

Arnould, R. and Grabowski, H. 1981. 'Auto Safety Regulation: An Analysis of Market Failure.' *Bell Journal of Economics* 12: 27–48.

Baram, M. 1980. 'Cost Benefit Analysis: An Inadequate Basis for Health, Safety, and Environmental Regulatory Decisionmaking.' *Ecology Law Quarterly* 8: 473–531.

Colantoni, C., Davis, and M. Swaminuthan. 1976. 'Imperfect Consumers and Welfare Comparisons of Policies Concerning Information and Regulation.' *Bell Journal of Economics* 2: 602–615.

Department of Transportation. 1984. *Guidance for Regulatory Evaluations: A Handbook for DOT Benefit-Cost Analysis.*

Environmental Protection Agency. 1983. *Guidelines for Performing Regulatory Impact Analysis.* EPA–230–01–84–003.

Ferguson, A. and E. LeVeen, eds. 1981. *The Benefits of Health and Safety Regulation.* Cambridge: Ballinger Publishing Co.

Fischhoff, B. 1977. 'Cost Benefit Analysis and the Art of Motorcycle Maintenance.' *Policy Sciences* 8: 177–202.

———, S. Lichtenstein, P. Slovic, R. Keeney, and S. Derby. 1982. *Approaches to Acceptable Risk: A Critical Guide.* Cambridge: Cambridge University Press.

Hershey, J., H. Kunreuther, and P. Schoemaker. 1982. 'Sources of Biases in Assessment Procedures for Utility Functions.' *Management Science* 28: 936–954.

Jones-Lee, M. W. 1976. *The Value of Life; An Economic Analysis.* London: Martin Robertson.

Kahneman, D. and A. Tversky. 1979. 'Prospect Theory: An Analysis of Decision Under Risk.' *Econometrica* 47 (2) 263–291.

Kleindorfer, P. 1982. 'Group Decision Making Methods for Evaluating Social and Technological Risks.' In *The Risk Analysis Controversy; An Institutional Perspective,* H. Kunreuther and E. Ley, eds. Berlin: Springer Verlag.

——— and H. Kunreuther. 1981. 'Descriptive and Prescriptive Aspects of Health and Safety Regulation.' In Ferguson and LeVeen 1981.

Kunreuther, H., J. Linnerooth, J. Lathrop, H. Atz, S. Macgill, C. Mandl, M. Schwarz, and M. Thompson. 1983. *Risk Analysis and Decision Processes: The Siting of LEG Facilities in Four Countries.* New York: Springer Verlag.

Lathrop, J. 1982. 'Evaluating Technological Risk: Perspective and Descriptive Perspectives.' In *The Risk Analysis Controversy: An Institutional Perspective,* H. Kunreuther and E. Ley, eds. Berlin: Springer Verlag.

Lave, L. 1981. *The Strategy of Social Regulation: Decision Frameworks for Policy.* Washington, D.C.: Brookings Institution.

———. 1982. *Quantitative Risk Assessment in Regulation.* Washington, D.C.: Brookings Institution.

Lawless, J. 1977. *Technology and Social Shock.* New Brunswick, N.J.: Rutgers University Press.

Majone, G. 1983. 'Reforming Standard-Setting.' *Journal of Policy Analysis and Management* 2(2): 285–290.

Mazur, A. 1982. *Dynamics of Technological Change.* New York: Sage.

Nelkin, D. 1981. 'Some Social and Political Dimensions of Nuclear Power: Examples From Three Mile Island.' *American Political Science Review* 75: 132–142.

Nuclear Regulatory Commission. 1983. *A Handbook for Value-Impact Assessment.* NUREG/CR–3568. Prepared by Pacific Northwest Laboratory.

———. 1984. *Regulatory Analysis Guidelines of the U.S. Nuclear Regulatory Commission.* NUREG/BR–0058 Revision 1.

O'Hare, M., L. Bacow, and D. Sanderson. 1983. *Facility Siting and Public Opposition.* New York: Van Nostrand–Reinhold.

Pauly, M., H. Kunreuther, and J. Vaupel. 1984. 'Public Protection Against Misperceived Risks: Insights from Positive Political Economy.' *Public Choice* 43: 45–64.

Peltzman, S. 1973. 'An Evaluation of Consumer Protection Legislation: The 1962 Drug Amendments.' *Journal of Political Economy* 81: 1049–1091.

Raiffa, H. 1982. *The Art and Science of Negotiation.* Cambridge, Mass.: Belknap Press of Harvard University Press.

The Royal Society. 1983. *Risk Assessment, A Study Group Report.* London: Royal Society.

Schoemaker, P. J. H. 1982. 'The Expected Utility Model: Its Variants, Purposes, Evidence, and Limitations.' *Journal of Economic Literature* 20: 529–563.

Slovic, P., B. Fischhoff, and S. Lichtenstein. 1983. 'Characterizing Perceived Risk.' In R. W. Kates and C. Hohenemser, eds., *Technological Hazard Management.* Cambridge, Mass.: Oelgeschlager, Gunn and Hain.

Snyder, B. 1984. 'EPA's Regulatory Impact Analysis Guidance Document and Its Application in the Decision-Making Process.'

Thaler, R. 1980. 'Toward a Positive Theory of Consumer Choice.' *Journal of Economic Behavior and Organization* 1(1): 39–60.

———. 1983. 'Using Mental Accounting in a Theory of Consumer Behavior.' Working Paper. Cornell University.

Tversky, A. and D. Kahneman. 1981. 'The Framing of Decisions and the Psychology of Choice.' *Science* 211: 354–358.
Walker, J. 1977. 'Setting the Agenda in the U.S. Senate: A Theory of Problem Selection.' *British Journal of Political Science* 7: 423–445.
Zeckhauser, R. and H. Leonard. 1983. 'Cost-Benefit Analysis Applied to Risks: Its Philosophy and Legitimacy.' Center for Philosophy and Public Policy Working Paper RC–6. University of Maryland.

Chapter 4

Conceptual Framework for Regulatory Benefits Assessment

Baruch Fischhoff and Louis Anthony Cox, Jr.

The present chapter provides guidance on procedural issues related to benefits assessment. The discussion considers how to exploit the entire set of benefits assessment techniques when commissioning analyses, defining their scope, and interpreting their results. It focuses on how to evaluate the definitiveness of an analysis and its contribution to policy making. Even a limited analysis of a difficult topic may be useful if its strengths and weaknesses are well understood. Even a polished analysis may be useless if it escapes peer review or avoids important but difficult issues.

Assessing the benefits of a regulation is an exercise in applied science. As with other research, the validity of a benefits assessment is constrained by the appropriateness of its problem definition and the truth of its empirical assumptions. In the case of benefits assessment, the problem definition determines just what is meant by "benefits," addressing issues such as which parties constitute legitimate stakeholders, whether distributional effects should be considered, and what standing should be given to the anxiety generated by a technology. The empirical assumptions underlying an assessment are assertions about how the world works that are presumed to be true for the sake of the assessment. They might address issues such as whether consumers understand the risks that they face in everyday life, what effect advertising has on marketplace decisions, and how precisely people's preferences can be measured. Using an inappropriate definition means asking the wrong question. Using inappropriate assumptions means getting the wrong answer.

As in any active science, controversies arise over the empirical assumptions. Investigators disagree, for example, about the extent to which consumers are "rational" (as defined by economic theory), firms maximize profits (as opposed to maximizing growth), and interviewees tell the truth about their preferences. As in any politicized domain, controversies also arise over the problem definition. Participants disagree, for example, over whether "benefits" should be defined as what people like or what paternalistic others believe to be good for them; whether people have entitlements, violation of which requires special dispensation; and whether to consider the benefits that some people derive from knowing that a natural resource has been preserved, even though they themselves will never use it. These controversies must be addressed in some way for an analysis to be conducted at all. They must be addressed effectively if it is to have any value.

Scope and Terms of the Analysis

Benefits assessment begins with a series of decisions that bound the analysis and specify its key terms. Together, these decisions provide an operational definition of what "benefit" means for the purpose of the analysis. Although they may seem technical and are often treated in passing, these decisions are the heart of an analysis. They express a social philosophy, elaborating what society holds to be important in a particular context. The ensuing analysis is "merely" an exercise in determining how well different policy options realize this philosophy. If the philosophy has not been interpreted, stated, and implemented appropriately, then the analysis becomes an exercise in futility.

In this section we discuss the issues that arise in defining "benefit," while the following section discusses the choice of methods for measuring it once that definition is in place.[1] Although most of that discussion concerns the validity of the empirical assumptions underlying those different assessment methods, a secondary focus is their implicit ethical positions about what definitions to use. These positions may be taken directly, as when a method endorses the status quo in society or precludes equity considerations; or indirectly, as when a method applies a diminishing label such as "intangible" to consequences that it is ill-suited to assess. In either case, the method dictates, rather than implements, social philosophy.[2]

The policy makers commissioning an assessment should be in the best position to determine what definition suits a particular regulatory context. If they abrogate this responsibility, then the task of identifying the appropriate social philosophy is left to the benefits assessor, who may not have adequate guidance. In such cases, the issues may never be faced explicitly. Instead, they might be left to unanalyzed "agency practice" or the customs of the assessor's discipline, which may be driven by other concerns (such as a preference for focusing on tractable and quantifiable issues). If definitions are not laid out and justified, then the customers of an assessment must first figure out what was done and then how meaningful it is.

For a governmental agency, the obvious source of guidance regarding these issues is its enabling legislation and the ensuing legal elaboration. For example, the agency and hence the analyst may be instructed to use "households" as the unit of observation, to reduce everything to current dollar equivalents, to concentrate on sensitive populations, to ignore psychological stress, to adopt a risk-neutral attitude, or to use "conservative" estimates. All too often, however, both the legislative mandate and the accrued interpretations are vague on crucial questions. Indeed, even so specific a directive as Executive Order 12291 (described in

4. A Framework for Regulatory Benefits Assessment

Chapters 2 and 3) allows enormous latitude in formulating an analysis. In such cases the policy makers themselves must decide what definition reflects the public interest. Facing ethical questions directly would forfeit the appearance of technicality that analysis sometimes can provide. It might, however, be a way to resolve questions that could come back to haunt the agency in the future. Whether this is a politically realistic risk might depend upon whether there is some hope that Congress and the Executive branch would accept the agency's leadership.

Diagnosing what problem has been solved is at least as important to the reviewer of an analysis as is evaluating how well it has been solved. The following section describes the definitional questions that need to be addressed in such a diagnosis. Because their answers depend upon the politics (broadly defined) of the problem being analyzed, no general prescriptions are given. Where those politics can be discerned, prescriptions follow quickly from this analysis. Where they cannot be discerned, a general definition of social philosophy must precede the specific definition of "benefit."

Variables in Benefits Assessment

Policy makers commission benefits assessments to help them make decisions; that is, to help them choose among alternative courses of action (including, typically, inaction). To make those decisions, they must: (a) identify the policy alternatives (or options) that could be adopted; (b) circumscribe the set of policy-relevant consequences that these alternatives could create; (c) estimate the magnitude of each alternative's consequences were it adopted; (d) evaluate the benefits (and costs) that affected individuals would derive from these consequences; (e) aggregate benefits across individuals. Defining the policy-making question is a precondition for commissioning any benefit assessment meant to serve it. For example, one cannot calculate the consequences of one particular policy without knowing the alternative policies that might come in its stead were it not adopted (and whose benefits would be foregone if it was). One cannot begin to assess and tally benefits without knowing which consequences and individuals fall within the agency's jurisdiction. The following discussion looks at the issues arising in each of these five aspects of specifying a decision problem. These are summarized in Table 4–1.

Table 4–1. Steps in problem definition

Identifying the set of policy options
  Specifying details of each option
  Determining the range of variation
  Assessing the uncertainty surrounding implementation
  Anticipating the stability of the situation following inaction
  Determining the legitimacy of creating new options arising during the analysis

Identifying the set of relevant consequences.
  Choosing consequences
    Scientific, legal, political, ethical grounds
    Public and private goods
  Specifying consequences
    Bounding in space
    Bounding in time
    Including higher-order consequences
    Including associated concern

Estimating the magnitude of consequences
  Assessing the uncertainty around estimates
  Determining risk assessor's attitude toward uncertainty
  Identifying deliberate bias in estimates
  Discerning the presuppositions in terms

Evaluating benefits for individuals
  Defining individuals
  Determining initial entitlements (willingness to pay versus willingness to accept)
  Identifying ultimate arbiter of benefit

Aggregating net benefits across individuals
  Looking for dominating alternatives (Pareto optimality)
  Exploring utilitarian solutions (potential Pareto improvement)
  Using group utility functions
  Resolving distributional inequities

Identifying the Set of Policy Options

Agencies make policies by issuing, interpreting, and applying regulations. In order to create viable regulations, agencies need to address at least the set of issues set forth in Table 4–2. Their success in doing so will determine how readily the standard can be understood and monitored, how likely it is to be overturned by courts or legislatures, how predictable a regulatory environment it will create, how quickly loopholes will emerge, and how faithfully the tradeoffs (between benefits, risks, and nonrisk costs) emerging from the operational standard will reflect the standard setter's intent.[3]

When the agency fails to specify its policy alternatives in sufficient detail,

4. A Framework for Regulatory Benefits Assessment

Table 4-2. Design variables in formulating operational standards

Range of application
> How broad is the category to which standard is applied?
> How extensively is the technology defined? (e.g., is the entire fuel cycle included)

Locus of the standard
> On the design (choice of technology, configuration of technology, operational procedures)
> On performance (permissible levels of initiating events, outcomes, consequences)

Mode of measurement
> Number of effects measured
> Aggregation rule
> Descriptive statistics
> Unit of observation
> Sampling scheme
> Directness of measure

Method of enforcement
> Dealing with uncertainty of interpretation
> Dealing with uncertainty of evidence
> Rigor of enforcement
> Margin of safety

Source: Fischhoff (1984)

the assessors must do so in order to get on with their job. If these imputed specifications differ from those eventually adopted, then the analyses take on an air of unreality. If they permanently fill the specification gap left by the policy makers, then the analysts have (perhaps inadvertently and unwillingly) helped shape policy.

When regulations are promulgated in a politicized arena, even the most careful specifications may not be a reliable predictor of the policy that will eventually emerge (once the lawyers, intervenors, and politicians get through with it). An assessor might, in fact, be advised to treat the policy itself (and not just its consequences) as an uncertain event whose exact configuration can only be estimated (see Brown 1975).

In addition to specifying its focal policy option, the commissioning agency must also identify the alternatives. A minimal option set would include two members: adopting the focal policy and rejecting it. The "benefits" of the policy then pertain to differences between the states of affairs obtained with and without it. Richer option sets might include minor variants on the focal point, which itself might be a minor variation on current practice. A significant design question is whether to include radically different possibilities, such as banning a regulated technology or totally deregulating it. Routinely considering such alternatives would encumber the analysts' task. However, it would also ensure that a broader perspective is maintained and help to highlight impacts common to all the minor variations.

Whatever options are considered, the same uncertainty surrounds their implementation as accompanies that of the focal policy. For example, the option of "doing nothing" contains no assurance that things will be left as they are. Regulatory inaction might lead to increased industry self-regulation, wanton exploitation of an apparent power vacuum, more court involvement (as parties turn there for redress), less court involvement (as the system stabilizes), reduced innovation (due to the rigidity of the regulatory system), or increased innovation (in a system that treats new products predictably).

Dealing with such hypothetical, subjunctive, or conditionally counterfactual alternative situations is difficult in any case (see Fischhoff 1980b). Yet some assumptions are needed if analysis is to continue. As a rule, those should be made for, rather than by, the analyst. The reviewer can then focus on how skillfully and faithfully the analyst considered the alternatives that the policy makers deemed relevant. Analysis can be a source of insight as well as of estimates, with the process revealing attractive new alternatives or variations on old ones which could be explored in subsequent iterations of the analysis. Letting analysts suggest alternatives may mean allowing them to usurp the policy maker's role. However, denying that possibility may unduly restrict their ability to provide guidance. Where such creativity is possible, it is hard to guarantee that the proposed policy is the best of all possible policies, or that the analysis has contrasted its benefits with those of the second-best alternative so as to reveal the opportunity cost of foregoing it.[4]

Identifying the Set of Relevant Consequences

Having determined what options are pertinent, the benefit assessor must then determine what difference they make; that is, what impacts will they have that might be construed as benefits. Table 4–3 shows one set of possible consequence dimensions (or "attributes") that might be considered when characterizing the attractiveness of policies. Winnowing this set down to those that will actually be considered has two distinct components: (a) roughly assessing whether any of the proposed policies has enough of an effect on each dimension for it to be worth bothering with, and (b) determining whether each remaining dimension has standing for that particular assessment. The first of these components requires scientific expertise and constitutes a prologue to the more precise consequence estimation needed for the next stage.

The need to look ahead here is just one of many instances in which the orderly set of steps presented in these sections is but an approximation of the

Table 4–3. Some possible dimensions of consequences for characterizing the attractiveness of options

Economic
  Compliance costs
  Market efficiency
    (e.g., monopolization,
    capital formation)
  Innovation
  Growth rate

Physical
  Death
  Genetic damage
  Injury
  Sickness

Ecological
  Species extinction
  Altered ecosystem balances
  Changed gene pools
  Habitat destruction

Political/ethical
  Centralization
  Inter- and intragenerational equity
  Personal freedom
  International relations
  Societal resilience

Psychological
  Worry, anxiety
  Confidence in the future
  Alienation

Source: Fischhoff *et al.* (1981)

actual process of problem definition. Another example is that the selection of options (Step 1) also requires some idea of what difference each will make (Step 3) and what differences need to be considered (Step 2). Moreover, as mentioned above, the process of analysis itself may reveal new options (or neglected consequences), prompting the iteration of steps while an analysis is under way.

The second of these components requires legal expertise to identify the consequences that an agency must weigh, political astuteness to identify those for which it will be held responsible, and social vision to identify those it is ethically appropriate to consider. Some of these consequences will be changes in the availability and prices of *private goods*, for which people compete in the marketplace (e.g., gasoline). Others are *public goods* (e.g., visibility, the protection afforded by levees), for which one person's consumption need not restrict that of another. The latter are often commonly desired consequences that could not be obtained without coordinated collective action, of the sort that an agency can provide (by initiating the action and levying the needed costs). These goods may include *paternalistic restrictions* on people's behavior (e.g., the protection afforded by mandatory seatbelts, the educational value of cigarette-package warnings) whose benefits may not be recognized by the recipients. Rather than providing the collective action needed to overcome institutional deficiencies, these measures provide coercive pressures to remedy individual deficiencies (e.g., misperceiving the risks of driving and ability to resist the social pressure to smoke).

Once the consequences have been identified, at least the following specifications must be made:

1. How to bound consequences in space. Government agencies, by definition, are responsible for problems within some limiting jurisdiction, perhaps a particular setting (e.g., workplaces), environmental medium (e.g., groundwater), or geographical locale. However, the effects of many technologies and the regulations governing them are less tidily bound. For example, restricting usage of a suspected carcinogen within the United States may also protect individuals abroad, or it may increase the export of that chemical; someone needs to decide whether "improving health" includes the changes elsewhere.[5] Restricting waterborne effluents may generally improve a substance's handling or it may prompt industrial processes with airborne releases; here, too, a specification is needed.[6]

2. How to bound consequences in time. The consequences of many technologies (and the policies meant to manage them) extend far into the future. For example, soil stabilization programs can protect a resource that benefits many future generations just as groundwater pollution can do the opposite. In such cases different time horizons can make a substantial difference in the magnitude of the consequences accruing from a project. A common analytical response is reducing future consequences to their *net present value*, namely, the amount of money that needs to be invested today at a specified rate of real interest (the discount rate) to accumulate to the amount of future benefit that is received or foregone. When the discount rate is high or the time period long, this procedure makes future effects negligible, allowing one to finesse specifying the time period. Unfortunately, this procedure makes little sense with health and safety consequences that cannot be banked at interest (although the economic effects of these changes might be). It also runs into difficulty with longer time periods, insofar as intervening generations can disrupt attempts to bequeath benefits to more distant generations (see Krutilla and Fisher 1975). The value of the rate (which can affect the assessment enormously) has been the subject of considerable unresolved debate, much of it over whether the fairly large values that are often used enshrine a "defective telescopic faculty" which is inappropriate for social decisions (even if people accept it in their individual decisions; see Pigou 1932 and Thaler and Shefrin 1981).[7] Moreover, some philosophers and economists have doubts about the ethical propriety of the entire discounting procedure (Schulze 1974; Mishan 1976; Goodin 1978; and Lind *et al.* 1982). When the discounting option cannot be used to make the time horizon relatively unimportant, the time frame must then be set on the basis of some first principles.

3. What higher-order consequences to consider. It is axiomatic that with interventions involving people, one thing leads to another.[8] For example, it has been claimed that people negate the safety benefits of accident-prevention measures by acting more recklessly in order to maintain a constant level

of risk.[9] It has also been argued that nuclear power regulations intended to improve safety may also bring economic benefit by reducing the rate of near-accidents which typically lead to widespread shutdowns (Evans and Hope 1982). When these higher-order consequences can be anticipated (at least on a probabilistic basis), the question arises of whether to include them as part of the direct consequences.

4. Whether to consider the concern associated with a consequence. Events that threaten people's health and safety can exact a toll even if they never happen. Concern over the risks of accidents, illnesses, and unemployment bother people even when they and their loved ones experience long, robust, and remunerated lives. Such concern may divert energy and financial resources from preferred pursuits; the accompanying stress can contribute to a variety of real health effects, especially when the threat seems uncontrollable (see Elliott and Eisdorfer 1983). Ironically, because the signs of stress are diffuse (e.g., reduced tolerance for marital tensions, aggravated cardiovascular problems), it is quite possible for them to be both unacceptably large and scientifically undetectable. A narrow way of recognizing changes in people's level of concern as a consequence of policies would be to include only demonstrable physiological effects; a broader basis would be that concern itself is a social cost whose reduction is a benefit.[10]

5. How to avoid double counting. In listing consequences it is important to count effects only once. For example, a safety regulation may reduce both a risk and the need for actions protecting against it (e.g., insurance purchases, worry, learning about the hazard). Although both savings are real, the benefit assessor must not count both the value of the property not destroyed and the insurance money not spent in replacing it. Or, the value of preventing (or achieving) higher-order consequences may be partially incorporated in the value attributed to their temporal predecessors. Either the shared value must be apportioned or the specification must choose the consequences of ultimate interest.

Estimating the Magnitude of Consequences

Determining how much of each consequence will be incurred is a matter of scientific investigation, treated in a companion volume.[11] Nonetheless, the conduct and reporting of scientific projects often involve political and ethical assumptions with implications for how their results are used in benefit assessments. Disentangling these issues of fact and value can also help policy makers give clearer directives to those scientists whose research is intended to serve benefits assessment.[12]

Benefits assessment cannot begin until there is some credible evidence that benefits exist. Creating such evidence requires deliberate scientific investigation. Whether it is undertaken reflects in part society's priorities as expressed in the funds available for particular research. However, it also reflects scientists' predilections for studying, say, cancer rather than neurological effects, or the pricing of houses rather than the pricing of visibility. Effects that cannot be substantiated tend to be neglected. Knowing this and the structure of science allow those who commission and review benefits assessments to ensure that poorly understood consequences are properly represented. Typically, this involves bracketing the best estimates of those consequences in appropriately broad uncertainty intervals. Doing so allows policy makers to know how much confidence to place in benefits assessments based on existing research, and scientists to know what additional research is needed (National Research Council 1983).

Scientists differ considerably in how they express such uncertainties.[13] Some give verbal qualifications; others append a statistical measure of the variability in observations; still others provide Bayesian confidence intervals. Summarizing the total uncertainty is often interpreted as a sign of weakness. Sciences that treat it explicitly may increase the vulnerability of policy makers attempting to justify their decisions. When the constraints of a situation restrict such candor, then there may be a preference for sciences and scientists that present a more confident veneer.

Scientists' attitudes toward uncertainty in their disciplines emerge in other ways as well: in readiness to treat hypotheses as "confirmed" (Mahoney 1979), in tolerance for internal debate regarding fundamental assumptions (Glass, McGaw, and Smith 1981; Hedges and Olkin 1980), in the rigor of procedures employed for aggregating results across studies, and for assessing the statistical significance of those studies (Page 1978, 1981). Knowing how scientists resolve these issues offers further cues for interpreting the products of their disciplines.

Other scientific practices affect the best estimates that these uncertainty expressions qualify. A common one in health and safety research is to provide "conservative estimates." That is, rather than reporting what they believe to be the truth, scientists attempt to err in the direction of safety. Although they may mean well and even express a widely shared prejudice, such scientists are preempting a role better left to policy makers. Moreover, without a consensual definition of how large these deliberate errors should be and where they should be introduced, one cannot ascertain the cumulative conservatism in overall consequence estimates. Perhaps policy makers should act even more conservatively; perhaps there has been overadjustment. In general, honest estimates are the best policy, however concerned scientists are as citizens.

Further mixing of science and politics can be found in the very definitions of scientific terms. To take but two examples:

1. A summary statistic is needed to express the magnitude of each consequence. For dimensions involving risk, a variety of summary statistics have

been used, including annual death toll, deaths per person exposed, deaths per hour of exposure, loss of life expectancy, loss of working days, and casualties per unit of production (Fischhoff, Watson, and Hope 1984). Although typically made on technical grounds, the choice of unit expresses an ethical position. For example, loss of life expectancy places a premium on early deaths which is absent from measures that treat all deaths equally; just counting fatalities expresses indifference to whether they come immediately after mishaps or following a latency period; including exposed individuals in a category ignores distinctions among them (e.g., between beneficiaries and nonbeneficiaries of a technology, workers and nonworkers, participants and nonparticipants).

2. In order to assess changes caused by an action, scientists need a standard of comparison. A natural baseline is the status quo, the situation that would obtain were the action not taken. As mentioned earlier, this requires factual assumptions regarding counterfactual states, often in a highly dynamic environment with the constant introduction of new products and regulations. Deciding which changes constitute benefits requires ethical assumptions regarding whether the status quo is "right." For example, smokers might consider restrictions on their freedom to smoke in public as a disbenefit that needs to be considered in assessing the consequences of smoking regulations; advocates of nonsmokers' rights might disagree (Fischhoff, Watson, and Hope 1984).

Policy makers should be able to specify the kind of scientific help they need. If they cannot, then they must be able to reinterpret the products of the science that they do get in order to have suitably unbiased and qualified estimates. Otherwise, policy makers' work may be distorted by the fads and prejudices of the science meant to serve them.

Evaluating the Benefits for Individuals

Once a consequence has been deemed a benefit (for the purpose of an analysis), a scheme is needed to translate changes in it into quantitative measures of benefit. Like its predecessors, this step requires a variety of technical specifications. Making these deliberately allows the commissioner of analyses to ensure that they reflect the desired social philosophy.

The first of these specifications is what constitutes an *individual* for the sake of the analysis. At one extreme, it is possible to treat society as a whole as the affected individual. At the other extreme, it is possible to treat each year (or day) in the life of each individual in society as the unit to which benefits and disbenefits accrue. Common intermediate positions are to look at the total benefits accruing to a "household" or "firm" over time.[14] In discussing aggregation across individuals, we will generally assume that the unit of analysis is the individual person, rather than any larger collective entity, so that social benefits are defined in terms of an aggregate of individual preferences.

Whatever unit of observation is chosen, the assumption is made that society is disinterested in the distribution of consequences within that unit. The ensuing analysis would be indifferent to situations in which all members of a household benefit equally and those in which more benefits accrue to parents than to their progeny. How controversial that assumption is depends upon the likelihood that members of the unit will exchange goods among themselves to rectify any perceived inequities. Such exchanges seem more likely in families, where parents worry about their children, than in firms where workers and shareholders have conflicting interests and little opportunity (or desire) for exchanges. Spontaneous rectification of perceived distributional inequities is virtually impossible when society as a whole is the unit of observation. Where it is possible to define individuals broadly, the obvious advantage to the analyst is avoiding the thorny problems of aggregating across individuals.

Individuals, however defined, can stand in two relations to the good or goods affected by the agency's decision: as current or potential possessors. Thus, the value that they attribute to its consequences might be defined as how much they would be willing to pay (WTP) to acquire the good or how much compensation they would be willing to accept (WTA) in order to relinquish it. These values can be further refined to: (a) maximum WTP for acquiring the good if not owned already, (b) maximum WTP for keeping the good if owned already, (c) minimum WTA for selling the good if owned, and (d) minimum WTA for foregoing purchase of the good.

From the viewpoint of economic theory, considerable attention has been given to the extent to which these values should differ from one another. A common conclusion is that for perfectly rational individuals in perfectly functioning markets, the perspective should make little difference, unless the value of the good approaches the individual's total assets.[15] Empirical investigations suggest, however, that people typically demand much more to give up something if they possess it than they are willing to pay to acquire it. This observation runs counter to the common-sense prediction that people grow weary of what they have (enshrined in such adages as "familiarity breeds contempt" and "the grass is always greener"). It might be traced to enhanced feelings of entitlement with WTA questions. If this difference proves robust, then the analyst needs to determine what constitutes the proper perspective for assessing benefits. A psychological basis for this choice would be whether actually possessing a good (WTA) gives individuals a better appraisal of its value than does hypothetically possessing it (WTP). An ethical basis would be whether the individual has a right to the good and, hence, should be compensated for its loss.[16]

A final specification is who constitutes the ultimate authority regarding a good's value to an individual. Although it might seem natural to rely on the individual's own tastes, some would argue that others at times know better. One justification for this (paternalistic) position is that people will overlook the virtue of actions that bring small benefits to each but large benefits to society as a whole. A more controversial justification is that people do not

4. A Framework for Regulatory Benefits Assessment

always know what is good for them, as when they insist on the right to Laetrile treatment or voluntary seatbelt usage. Those making such claims must rely on their own tastes for assessing the magnitude of the benefits brought by the paternalistic restrictions that they would impose, insofar as the traditional sources of guidance (what people say and what people do) all assume the sovereignty of people's preferences.

Aggregating Net Benefits across Individuals

Having determined how good an action is for the various "individuals" in society, the benefit assessor must then determine how good it is for society as a whole. An ideal situation, for both society and the analysts, arises when the action would leave at least some individuals better off and no one worse off. The Pareto criterion, or principle of unanimity, which calls for accepting such options, is widely accepted among decision theorists.[17] Unfortunately, it does not widely occur in reality. As soon as many individuals are involved, there are likely to be losers as well as winners. For example, many regulations have widely distributed benefits, with costs focused on a relatively small group (or industry). In such cases the Pareto criterion offers no guidance, and some direct comparison among the benefits accruing to different individuals is required.

For the problem of interindividual comparisons, there is no generally accepted solution. A straightforward but flawed approach is to see whether there are more winners or more losers. Obviously this procedure ignores the magnitude of the gains and losses that each suffers (or assumes that they are roughly equal), so that many small winners could outweigh a few very large losers. It allows for a tyranny of the majority, as well as for situations in which a few big winners can bribe others into supporting a measure by adding a little something for them, while leaving a weighty burden for a minority.

A more sophisticated solution would consider the magnitudes of the gains and losses. In one version of the utilitarian rules following from this perspective, the most attractive action is that producing the greatest preponderance of gains and losses as represented by the arithmetic sum of net consequences across all individuals. A technical limitation of such rules is that they require reducing the net consequences for all individuals to a common numerical measure. As discussed below, this may not be possible, even for those consequences that can be reliably evaluated in monetary terms. An ethical limitation is that the rule is apathetic to distributive equity considerations, the way in which some individuals benefit at others' expense.

A refinement that attempts to deal with this ethical problem is the *potential Pareto improvement criterion* (often known by the name Kaldor-Hicks). It endorses policies whose gains and losses could, in principle, be redistributed so that no individual would be worse off (and at least one would be better off). Unlike the crude utilitarianism described above, this rule asks whether the winners would be willing to pay more in order to buy off the losers than the losers would demand in compensation (before they would accede to the policy). Although it acknowledges the losers' situation, this rule does not require the winners actually to pay compensation or the losers actually to give consent. It ignores the losers' possible resentment at being (only) no worse off while the winners are made better off. It effectively enshrines current inequalities by giving disproportionate influence to individuals with more resources, insofar as they can pay more to secure policies that they favor and demand more to accept policies that they oppose. It also assumes that everything has its price, leaving no role in pricing for questions of fundamental rights and entitlements. It would not, for example, accept the logic of the considerations motivating the Occupational Safety and Health Agency regulations forbidding "excessively" hazardous work conditions, even if employers can offer wages sufficient to get workers to accept the jobs.[18]

Some of these difficulties can be avoided by replacing monetary evaluations with a general-purpose evaluative scale, most commonly known as *utility*.[19] A typical utility assessment procedure would have individuals assign values of 0 and 1 to the outcomes they prefer least and most and intermediate values to other outcomes by reference to these two end points. As described in Chapter 5, sophisticated methods exist for formally structuring the elicitation of such utilities.[20] Such procedures facilitate assigning values to "goods" that are not traded at all, such as the aesthetic properties of landscapes. By neutralizing the importance of current asset positions, such procedures also reduce the influence of existing inequalities in the distribution of resources.[21]

Individual utility measures may also provide the basis for a group utility function. Constructing such a function requires two strong empirical assumptions; namely, that all individuals in a group are fully "rational" (as defined by economic theory) and that all individuals' utilities can be measured up to interval scale precision (so that equal differences in utility scores can be interpreted as equivalent differences in strength of preference). It also requires the ethical assumption that when every group member is indifferent when faced with two alternatives, the group as a whole should be indifferent.[22] Under these conditions, the group's utility for an action should correspond to the sum of the individuals' utilities, weighted by the importance assigned to the individuals. This is also the only group utility function satisfying the Pareto condition and a condition of continuity (see Chapter 5). The weighting could be equal for all individuals, or it could be sensitive to need, merit, or power; it could reiterate society's distribution of resources or promote a redistribution. Thus, although group utility functions can make the same controversial assumption as monetary utilitarianism regarding society's current distribution of assets, they can also make alternative controversial assumptions. Controversy is inherent in either procedure; explicit weighting only makes it more apparent.

As mentioned above, additional equity assumptions are embedded in the definition of "individual," with distributional questions being ignored for those within a unit of observation. It is assumed that group members who lose personally are content as long as the group as a whole gains. Such equanimity may mean that they are receiving side payments from the winners or that they have altruistic feelings regarding those winners. Or, it may mean that group membership works out acceptably over the long run, so that even though they are losers for one event, they are winners for others. At the societal level, this perspective could express itself in thoughts such as, "We're all in this energy-consumption business together. I'm willing to live near this nuclear power plant, if you're willing to live downstream from that dam." In effect, the unit of observation would then be expanded to include multiple events as well as multiple individuals.

Such complex balancing acts are probably beyond most individuals' range of explicit considerations and the analytical capacity of any mechanical rule. They seem more suited to the more informal deliberations of policy makers searching for some passable solution. Rather than attempting to compute a group utility function as a way to treat the aggregation problem, they are more likely to examine the utilitarian solution, then adjust it to rectify the most egregious inequities.

A hallmark of responsible benefit (or risk) assessment is the conduct of sensitivity analyses, examining the effect on policy conclusions of varying parameter estimates along the range of plausible values. Doing so gives systematic representation to the uncertainty in estimates and relieves policy makers of the mental arithmetic of deriving those solutions by themselves. The uncertainties regarding aggregation have to do with alternative rules and not alternative numerical values. Sensitivity analysis here means computing the solutions arising from the use of alternative rules. Doing so will show the public what assumptions lead to the policy eventually adopted.

Empirical Assumptions in Benefits Assessment Methods

Role of Factual Assertions

Beginning benefits assessment with a detailed specification puts the problem in the driver's seat. It focuses method selection on the ability of competing methods to capture the values that people ascribe to the particular consequences being considered and the societal values embodied in the problem definition. A method with such descriptive validity must be able to discern people's true preference within their observable behavior (either words or action). The following two sections discuss some of the factors that can affect the expression of those preferences. All methods make some assumptions regarding the prevalence and potency of these factors. Knowing which of these factors are at work in particular situations allows choice of the most suitable method(s) and the qualification of the results obtained with less suitable ones. These factors can be divided into aspects of how individuals and society function.

Nature of Individual Behavior

An extreme but common assumption regarding the behavior of individuals is that they are fully optimizing in their decision making; that is, that they have (1) a full understanding of whatever can be known regarding all possible contingencies, (2) sharply articulated preferences among all possible options, and (3) the ability to combine all these considerations in accordance with the tenets of rational decision making. Empirical study (and common sense) suggest reasons to qualify each of these assumptions.[23]

It is unreasonable to expect people to be fully informed regarding all of life's challenges, particularly for those that are novel, infrequent, and cloaked in technical language. As might be expected, lay people have frequently been found to be relatively ignorant or misinformed regarding the extent of many contemporary risks to health and safety (see Slovic, Fischhoff, and Lichtenstein 1979, 1980, 1984). Even workers in hazardous industries seem only crudely aware of the risks that they face on a routine basis (see Chapter 7), a state of affairs that may be partially causing the current pressure for right-to-know statutes among chemical industry workers. Where the sources of people's inaccurate beliefs can be traced, they seem relatively defensible, in the sense of being signs of ignorance rather than of stupidity. People seem to know the most about the events that are most germane to them and for which accurate information is most readily accessible. Where they must guess at facts, they seem to rely on judgmental rules of thumb that are sensible but highly imperfect. These psychological processes leave people's perceptions relatively vulnerable to manipulation (whether deliberate or inadvertent).[24]

A common concession to human limitations in the realm of preferences is recognition of the fact that people are not capable of infinitely fine evaluative discriminations in most stimulus domains. Theoreticians have developed sophisticated schemes for characterizing people's preferences as a function of their degree of precision (Fishburn 1982; Krantz et al. 1971). One common description of these characterizations is in terms of level of measurement. In order of increasing precision, these levels include: the ability to make *binary comparisons* with some standard (such as the status quo); the ability to *partially order* objects by making (transitive) comparisons between the members if some, but not all, pairs; the ability to *rank order* all objects; the ability to order the *differences* (or intervals) between objects; the ability to assign *cardinal* measures of worth such that the ratios of numbers are meaningful.

These theories prescribe empirical tests for determining whether people's preferences are sufficiently well articulated to be represented at each level of measurement. Knowing the appropriate level allows benefit assessors to have realistic expectations regarding the individuals whom they observe. Individuals who can just judge whether they prefer alternatives to the status quo will be hard put to assign cardinal dollar values to those alternatives.

4. A Framework for Regulatory Benefits Assessment

Requiring greater precision than people can muster is likely to produce inconsistencies and measurement artifacts. The precision of individual preferences obviously constrains the precision that can be expected in societal benefit assessments.

The level-of-measurement approach assumes that people have orderly preferences with a degree of precision reflecting how distinct the alternatives are and how much they care about them. However, a growing body of evidence (and perhaps common sense) indicates that people do not have orderly preferences regarding many sets of alternatives that are both diverse and important (see Fischhoff, Slovic, and Lichtenstein 1980; Hogarth 1982; Kahneman and Tversky 1979; Rokeach 1973; Tversky and Kaheman 1981). Rather, it seems that they have pieces of preferences, in the form of basic values that are strongly held and intimately related to the topic, but not integrated into a coherent preference. If forced to make an evaluative judgment, they engage in an excercise in inference, deducing the implications of their related values and beliefs. That inferential process is most likely to produce reliable conclusions when individuals have the opportunity for thoughtful rumination over the issues, for consultation with informed others, and for firsthand experience with the alternatives (and their consequences) to serve as a check for the conclusions derived intellectually (National Research Council 1982; Turner, Martin, and De Maio 1985).

Convergence on a stable position is least likely when people are unfamiliar with the terms in which issues are formulated (e.g., social discount rates, minuscule probabilities, megadeaths); when there are unresolved conflicts among their basic values (e.g., a special aversion to catastrophic losses of life and a realization that an accident with 500 casualties is little more moving than one with 300; the belief that freedom of speech is inviolate but should be denied to authoritarian movements); when they occupy different roles in life (e.g., parents, workers, children) which evoke clear-cut but inconsistent values; when the problem juxtaposes very diverse consequences (e.g., requiring a tradeoff between dyeing one's hair now and a vague, minute increase in the probability of cancer 20 years hence).

In such situations, subtle variations in how problems are posed can effect marked differences in the values expressed. For example, most people prefer a gamble with a 0.25 chance of losing $200 to a "sure loss" of $50, but would prefer to pay an insurance premium of $50 to guarantee against the loss; the relative attractiveness of two gambles may be different when people are asked how much they will pay to play each than when they are asked which they would prefer to play; a proposed public health medical intervention may seem more attractive when described in terms of the lives that would be saved, than in terms of the lives that would be lost. To an observer, these vacillations may appear to be the product of labile or erratic preferences. However, they may be better seen as cases in which but a portion of people's values are evoked by the problem description facing them. Such incomplete evocation and integration of basic values can occur in any setting, including laboratory experiments, field surveys, and real-life marketplace decisions.

A substantial body of theory has evolved to account for these phenomena.[25] Most of these theories describe the basic processes governing people's decision making in terms that are incompatible with the model of rational decision making. For example, the model in Tversky's *Elimination by Aspects* (1972) has individuals comparing alternatives one aspect (or consequence) at a time, in some order, until one alternative shows some superiority; it is then chosen, without any assessment of the overall attractiveness of the alternatives. "Satisficing" processes (Simon 1956, 1978) deviate from the traditional view of rationality, as do those decisions that rely on unanalyzed habit or tradition or most of those described in artificial intelligence models of expert human inference (Fox 1980).

With this research has come a general understanding of the conditions under which people are more or less likely to behave like the optimizing hedonists described in models of rationality. Some of the barriers include overwhelming complexity and detail (Goldberg 1968; Meehl 1954; Miller 1956); the absence of prompt, unambiguous feedback regarding the adequacy of previous decisions (as might happen when the passage of time allows intervening events or the biases of hindsight to blur the message) (Einhorn and Hogarth 1978; Fischhoff 1975); changes in the criteria for evaluating decisions between the time when they are made and the time they are judged (Bell 1982; Fischhoff 1980a); the presence of strong normative rules, representing societal traditions or ingrained individual habits that circumvent the need for analytic decision making (by providing ready-made responses).

Diagnosis of the extent to which these conditions obtain allows one to anticipate when behavior will be nonrational. An "error theory" would then be needed to determine the extent to which these deviations from the processes of optimal decision making will result in suboptimal decisions (Dawes 1979; von Winterfeldt and Edwards 1982; Fischer 1977). Descriptive theories of decision making might provide substantive predictions regarding which decisions people will actually make.[26]

Nature of Societal Behavior

A common and extreme assumption about the nature of society is that it offers people opportunities for acting freely on their preferences. That is, they have a range of options from which to choose, suitable information upon which to base that choice, an understanding of how their preferences apply, and the chance to select options in their own best interests. In that case, it should be possible to observe an expression of those preferences that can be exploited as an input to benefits assessment. These are the conditions that characterize a free-market economy. As a result, any restraint on trade also restrains people's ability to express their values; it is thus bad for benefit assessment, as well as being bad for society. The threats to the market and hence to preference expressions lie in factors that reduce the range of options offered: the availability of information to actors in the marketplace, the ability of those actors to identify the options that are in their own best interests, and the participation of individuals in the decisions that affect them.

The range of options may be unduly restricted for several reasons. Monopoly, monopsony, excessive concentration, and other forms of market thinness may discourage the development of products (such as those that would weaken the authority of unions). Lack of imagination or mistaken beliefs about consumer preferences (e.g., the feeling that safety does not sell) may inhibit the creation of desired options. Regulation may discourage innovation (and the options that it brings). A free market cannot, in principle, create options that require coordinated action. One result of this limitation is the "tragedy of the commons," whereby uncoordinated individually rational decisions deplete a shared resource (such as grazing land, a fishery, an airshed), leaving everyone worse off (see Dawes, Delay, and Chaplin 1974; Hardin 1968; Schelling 1978). Whether collective action options, such as regulation, licensing, and rationing are created to supplement market options may depend upon factors such as whether those concerned are sufficiently concentrated geographically and informed politically to act effectively.

The information offered to people may be restricted or distorted for several reasons. Where knowledge is power, those with knowledge may restrict its flow, perhaps releasing it selectively, so as to create particular impressions, if not engaging in outright misrepresentation. Where knowledge is technical, lay people may have little access to it unless scientists make dissemination a priority. Where people exaggerate the extent of their own knowledge, they may not recognize the need to learn (see Fischhoff 1982; Lichtenstein, Fischhoff, and Phillips 1982). Where there are long latency periods between events and consequences or events are multiply caused, thereby complicating the attribution of responsibility, good risk information may be hard to come by for both experts and lay people.

Society's restrictions of its citizens' ability to identify the best option are more subtle. A deep constraint is the distribution of political and economic resources within society; people may not seriously consider options that are beyond their means, however attractive they might be. These are facts of life (perhaps the ultimate facts of life). They are important here in that they suggest interpreting people's choices as "the options that seem best to them, among those that they are able to consider."

Societal constraints of a more psychological character come from attempts (by advertisers, public relations specialists, political activists, and so on) to tell people what their options are and choices should be. If successful, such coercion can notably forestall deliberative decision making but also confuse people's attempts to discern their own values and interests. An additional societal imposition is the need to make diverse decisions under frequent time pressure. This fact of modern life, too, can limit the extent to which people's actions mirror their preferences. Being forced to have some opinion on many issues may ensure having thoughtful positions on none.

Finally, the market can fail by affording people no say in decisions that affect them. When the result is the imposition of involuntary and uncompensated losses (or the risk of losses), those decisions are said to create *externalities*. A single factory might produce several examples of externalities if its smokestacks soiled the clothes, effluents polluted the groundwater, and physical plant disrupted the landscape for individuals unable to negotiate an acceptable deal for themselves. In such cases, the risks they bear and the benefits that they derive from them may say more about their social power than about their tastes.

Diagnosing how severely these societal constraints impinge on people's ability to make decisions reflecting their own values requires an analysis of their situation and the sensitivity of their decision-making processes to it. That analysis provides a basis for knowing how heavily to rely on benefit assessment methods that assume an idealized, free-market environment. A useful adjunct would be a theoretical understanding of how sensitive market outcomes are to both structural flaws (e.g., missing options) and relative imperfections (e.g., restricted information). A convenient summary for this diagnosis would be that it asks whether the observed behavior was obtained under conditions of *informed consent*.

Choosing a Method for Benefits Assessment

Assuming that people's preferences are the ultimate source of guidance for the policy decisions meant to govern their lives, there are two natural places to look for insight: in what people say and in what they do. Methods relying on the former consider *expressed* (or announced) *preferences*; methods relying on the latter, examine *revealed preferences*. Each makes certain general assumptions regarding the nature of individual and societal behavior, whose validity determines their suitability for particular applications.

Expressed Preferences

The most straightforward way to find out what people prefer is to ask them. The asking can be done at the level of overall assessments ("do you favor . . .?"), statements of principle ("should our society be risk adverse regarding . . .?"), or detailed tradeoffs ("how much of a monetary sacrifice would you make in order to ensure . . .?"). The collection vehicle could be public opinion polls,[27] comments solicited at public hearings (Mazur 1973;

4. A Framework for Regulatory Benefits Assessment

Nelkin 1984), or detailed interviews conducted by decision analysts (Keeney 1980; Janis 1982). The advantages of these procedures are that they are current (in the sense of capturing today's values); specifiable and controllable (in the sense of allowing one to ask the precise question that interests policy makers), sensitive (in the sense of allowing expressions of the need for change); direct (in the sense of looking at the stated preferences themselves and not their application to some specific decision problems); superficially simple (in the sense that you just ask people questions); and politically appealing (in the sense that they let the people speak).

The preceding analysis suggests, however, a number of conditions that must be met if expressed preference procedures are to meet their potential. One is that the question asked be the precise one needed for policy making (e.g., "how much should you be paid in order to incur a 10 percent increase in your annual probability of an injury sufficiently severe to require at least one day of hospitalization, but not involving permanent disability"), rather than an ill-defined one such as "do you favor nuclear power?" or "is your job too risky?"[28] An obvious strategy for meeting this challenge is to follow the details of the problem definition closely when formulating the questions put to the respondents. However, it may be ineffective if the full specification is so complex and unfamiliar as to pose an overwhelming inferential task for respondents. To avoid the incompletely considered, and potentially labile, responses that might then arise, one must either adjust the questions to the respondents or the respondents to the questions. The former requires an empirically grounded understanding of what issues people have considered and how they have thought about them, allowing one to focus the interview on those areas in which people have articulated beliefs, to provide needed elaboration (e.g., regarding the intended meaning of terms), and to avoid repeating details that correspond to respondent's default assumptions (and should, therefore, go without saying).

If the gap between the policy maker's questions and the respondent's answers is too great to be bridged in a standard interviewing session, then it may be necessary to either simplify the question or complicate the session. A structured form of simplification is offered by techniques, such as multi-attribute utility theory,[29] which decompose complex questions into more manageable components, each of which considers a subsidiary evaluation issue. The structuring of these questions allows their recomposition into overall benefit assessments, which are interpreted as representing the summary judgments that respondents would have produced had they had unlimited mental computational capacity. The price paid for this potential is the need to answer large numbers of simple but formal, precise, and hypothetical questions (see Chapter 6).

Where policy makers do not require overall assessments but general guidance regarding the kinds of benefits to consider or the weights to assign them, then the needed simplification might be accomplished by soliciting responses on basic policy issues (e.g., should society be particularly averse to losing large numbers of lives in single accidents as opposed to losing the same number in single accidents? Slovic, Lichtenstein, and Fischhoff 1984). Is any (nonzero) discount rate appropriate for threats to health and safety? (Schulze 1974; Mishan 1976; Goodin 1978; Lind et al. 1982).

Yet another question-simplification strategy leaves the stimuli intact but makes the response mode less demanding. That is, instead of requiring explicit quantitative evaluations of alternative policy options or possible consequences, ask people just to rank order them, or to choose between pairs. There are elaborate methodologies for doing so, each of which requires responses at relatively low levels of measurement but is capable of deriving relatively high level measurements from them (if certain conditions are met) (Coombs 1964; Krantz et al. 1971; Fishburn 1982).

Where it proves impossible to bring the question "down" to the respondent, there still may be some opportunity to bring the respondent "up" to the question. Ways of enabling respondents to realize their latent capability for more reflective evaluation include: letting them talk about the issues, including them in focused group discussions, suggesting alternative perspectives (for their consideration), and giving them time to think about their answers.

Although common in some contexts (e.g., marketing), these procedures are seldom components of survey research. One source of resistance is the desire to ensure statistically reliable results, which requires large numbers of carefully selected individuals – conditions that are hard to meet with more intensive procedures. A second source is the fear of influencing respondents should the interviewer abandon an impassive, nondirective demeanor. Where respondents' beliefs are incompletely articulated, however, the danger of influencing those beliefs must be weighed against the danger of improperly representing them. Only a portion of respondents' values may be elicited by a procedure that asks but a few questions; then strongly held fundamental values may emerge as labile and untrustworthy preferences on specific questions. An intermediate position provides radically different perspectives that respondents might wish to consider, but then leaves it to them to make whatever integration seems appropriate.

Assuming that respondents understand the question in an expressed preferences study, the validity of their responses then depends upon whether they are telling the truth. Deviations from the truth could come from several sources. One is a *strategic response*: deliberately misrepresenting one's views so as to influence the results of the analysis (e.g., understating one's willingness to pay taxes for toxic water cleanup in hopes of increasing the burden on industry). Such possibilities have been studied intensively in recent years, with the evidence suggesting that the problem exists more in theory than in practice. Indeed, rather than lying, strongly motivated respondents may choose not to participate at all in interviews that do not allow full expression of their complex beliefs (Brookshire, Ives, and Schulze 1976). Another kind of misrepresentation, requiring much less sophistication, occurs when respondents provide the answers designed to make them look good to the interviewer. Avoiding such *social desirability effects* is a focal topic in survey research methodology, with a common response being the use of impassive, even nondescript interviewers attempting to avoid any hint at what the answer should be (Payne 1952; Bradburn and Sudman 1979).

An apparent symptom of misrepresentation is the observation of behavior that differs from what one would expect on the basis of people's expressed preferences (e.g., they say that they favor integration but move to the suburbs along with everyone else when the "neighborhood begins to change"). Such *attitude-behavior inconsistency* has been actively studied for many years (e.g., Schuman and Johnson 1976; Ajzen and Fishbein 1977), one conclusion being that individuals consider themselves to be much more consistent than do those who observe their behavior. People often see reasons for their actions that escape observers (e.g., "being against racism does not mean being insensitive to property values"). The need for inferring the causes of behavior can be reduced by making the attitude question similar to the behavioral question. Doing so reduces inconsistency (although in the extreme, it means asking people to predict what they will do).

A form of inadvertent misrepresentation occurs when people do not understand the reasons for their behavior. That is, they know what they like but are misinformed about why they like it (Nisbett and Wilson 1977). Although this state of affairs may seem implausible, it appears to be moderately common with more complex cognitive processes. Two contributing causes are the intrusion of widely accepted (but inaccurate) explanations for why people do things and the speed with which people execute familiar inferences (Goldberg 1968; Ericsson and Simon 1980). One approach to improving introspection is to slow down the evaluation process by asking people to discuss evaluative considerations as they arise (Newell and Simon 1972; Svenson 1979; Ericsson and Simon 1984). A second methodological response is to have people make many summary judgments in their normal intuitive way, after which they are shown a model describing the factors that determined their preferences. By iteratively adjusting the model and the judgments, people may be able to derive a satisfactory analytic representation of their evaluation philosophy (Hammond and Summers 1972). A less elaborate version of the same strategy simply reflects back to respondents the assumptions that apparently underlie their judgments. Doing so means achieving the sort of helping relationships that Socrates is supposed to have had with his disciples rather than the sort of bullying that seems to have actually characterized them.

Although troublesome, these possibilities are fairly well understood. The various disciplines engaged in expressed preferences have conducted methodological studies into the elicitation problems that most concern them. These provide a basis for anticipating how well people's values are articulated for various issues, developing appropriate elicitation procedures, and assessing the residual imprecision in the resulting responses.

Revealed Preferences

The alternative to words is action. Proponents of revealed preference approaches assume that people's overt actions can be interpreted to reveal the preferences that motivated them. The great attraction of such procedures is that they are based on real acts, whose consequences presumably are weightier than those derived from an intelligently conducted interview. It is hoped, for example, that in actual decisions people will sharpen their feelings regarding what they really want (rather than be overcome by time constraints, social pressures, and emotional concern over the consequences). Their proponents see virtue in the fact that these procedures, unlike expressed preferences, are relatively insensitive to changes in public values (which might be unrealistic even if they are not ephemeral).

By focusing on current real decisions, these procedures are strongly anchored in the status quo. It is today's world, with today's constraints, that conditions the behavior that is observed. If today's society restricts people's ability to act in ways that express their fundamental values, then these procedures lose credibility. As a result, revealed preferences are particularly sensitive to the issues raised in the dicussion above on the nature of society (whereas expressed preferences, at least in principle, allow people to raise themselves above today's reality). Thus, if one feels that advertising, or regulation, or monopoly pressures have distorted contemporary evaluations of some products or consequences, then revealing those values does not yield a guide to true worth. Using those values for policy making would enshrine today's imperfections (or inequities) in tomorrow's world.

The commitment to actual behavior also makes these procedures particularly vulnerable to deviations from optimality. A much smaller set of inferences separates people's values from their expression of preferences than from their actual decision-making behavior. Thus, for example, it is difficult enough to determine hypothetically how much compensation one would demand to accept a risk of magnitude X in one's job. Implementing that policy requires an analysis of the complex, often incommensurable tradeoffs in the options actually offered. For the choice to be a true reflection of one's values also requires that suitable options be available and that their consequences be accurately perceived. If these conditions of informed consent are not met, then interpretation of wage-risk relationships becomes tenuous. Workers may coerce their employer into compensating them for imagined risks, or be coerced into accepting minimal compensation by employers cognizant of a depressed job market, or be exposed to risks beyond compensation.

The most common kind of revealed-preference analysis, which is also the most common kind of economic analysis, interprets marketplace prices as indicating the true values of goods. If a good is not actually traded, a value for it might be inferred indirectly by conceptualizing it as a bundle of attributes and then seeking the prices that the market assigns to each attribute individually (with their sum giving the worth of the object as a

4. A Framework for Regulatory Benefits Assessment

whole). If an attribute is not traded individually, then inferential procedures might be able to discern its role in determining the price paid for goods offering it.

A typical revealed-preference problem might be determining the value of safety to workers by studying the wages earned for jobs varying in risks (see Chapter 7). If jobs were identical in all respects other than risk level, then the "safety premium" would be readily calculated. However, identical jobs are scarce when one considers all the factors that can affect wages (e.g., comfort, unionization, security, local unemployment, co-workers, training requirements, predominant sex). Riskier jobs might even be more poorly paid if they were also entry-level jobs or in economically depressed areas. Nonetheless, their wages might still be more than they would be were their risk level lower. In order to investigate this possibility, analysts might use a statistical procedure such as multiple regression, which examines whether knowing the risk level of jobs facilitates predicting their pay level after holding other relevant factors constant (see Chapters 6 and 8 for details).

Such analyses require attention to a great number of technical details, such as the consequences that are included as predictor variables in the regression equation, the correlations among these variables,[30] the ratio of sample size to number of variables, and the reliability with which the different variables are measured (see, e.g., Cohen and Cohen 1975). Once the analyses have been done, additional care is needed in generalizing their results. Strictly speaking, they pertain only to other items in the same population (with the same intercorrelations, ranges of values, etc.). As one goes farther from that population, the conclusions become less useful. Thus, if the study just described concerned workers in a depressed rural area, one might be reluctant to use the value of safety that it derives in a benefit assessment regarding workers in a prosperous suburban area, or even for the rural workers after unionization. One would further qualify any generalization with an assessment of the extent to which the conditions of a free market and informed consent characterized the decisions upon which it is based (e.g., did the workers involved have other job opportunities? Would refusing hazardous work imperil their seniority? Did they have an accurate appraisal of the risks and of their habituation to them?).

These procedures rest on a well-developed theoretical foundation describing why (assuming a free market, optimal decision making, and informed consent) prices should reveal the values that people ascribe to things. The same general thought has been applied heuristically in various schemes designed to discern the values revealed in decisions (ostensibly) taken by society as a whole or by individuals under less tightly constrained conditions. These include attempts to see what benefits society demands for tolerating the risks of different technologies (Starr 1969; Otway and Cohen 1975), what risks people seem to accept in their everyday lives (Cohen and Lee 1979; Wilson 1979), and what levels of technological risk escape further regulation (Nuclear Regulatory Commission 1982; Fischhoff 1983). These attempts are typically quite ad hoc, with no detailed methodology specifying how they should be conducted. The implicit underlying theory assumes, in effect, that whatever is, is right and that present arrangements are an appropriate basis for future policies. Thus, they can only guide future decisions if one assumes that society as a whole currently gets what it wants, even with regard to regulated industries, unregulated semi-monopolies, and poorly understood new technologies. Extracting useful information from such studies requires a detailed assessment of the procedures they use, the existing reality that they endorse, and the kinds of behavior that they consider.[31]

A more articulated theory for modeling imperfect market situations can be found in the estimation of *shadow prices* (McKean 1972). If it can be assumed that society as a whole manages to maximize the social welfare of the individual consumption and production levels of all (and not just market) goods, within technological and resource constraints, then it can be argued that the relative value of consequences is revealed by the rate at which consequences (or goods or attributes) are substituted for one another. Actually deriving these values is complicated by the technical difficulties of finding substitution-rate information, dealing with a dynamic economy, and incorporating uncertainties, as well as by the ethical question of defining social welfare.

Ascertaining the validity of the theory underlying approaches assuming optimality has often proven difficult, for what can best be described as philosophical reasons. Some investigators find it implausible that people would do anything other than optimize their own best interests when making decisions, maintaining that our society would not be functioning so well were it not for this decision-making ability. These investigators see their role as discerning what people are trying to optimize (i.e., the values that they ascribe to various consequences).

The contrary position argues that this belief in optimality is tautological, in that one can always find something that people could be construed as attempting to optimize. Looking at how decisions are actually made shows that they are threatened by all the problems that can afflict expressed preferences. Thus, for example, consumers may make suboptimal choices because a good is marketed in a way that evokes only a portion of their values, or because they unwittingly exaggerate their ability to control its risks.

Because of the philosophical differences between these positions, relatively little is known about the general sensitivity to deviations from optimality of conclusions drawn from analyses that assume optimality. The consumer of such analyses is left to discern how far conditions deviate from optimal decision making by informed individuals in an unconstrained marketplace and then how far those deviations threaten the conclusions of their analyses.

4. A Framework for Regulatory Benefits Assessment 77

Using Benefits Assessments

The chapters in this volume constitute a consumer's guide to benefits assessment, describing what techniques are available, along with their strengths and weaknesses. They should help policy makers to specify the analyst's mandate, order suitable techniques, and interpret their results. Of course, benefits assessments are commissioned not so much because of their inherent interest but because of their potential contribution to policy making. Reviewing this volume and chapter from this perspective suggests a number of general conclusions regarding how the process of benefits assessment can be integrated into the process of policy making and how the conclusions of assessments can be responsibly incorporated in policies.

Managing Benefits Assessments

The need to conduct some benefits assessments is a foregone conclusion in the setting of many policies. The key decisions are when, which, and how much assessment are desirable. A detailed problem definition not only tells the analyst what work to do but the policy makers what analysis to choose. The definition can show the political values needed in a technique consonant with the purposes of the analysis. For example, if a policy question arises from dissatisfaction with current expenditures on consumers' safety, then it is inappropriate to rely on a technique that endorses the status quo. Having a precise definition can also show where the greatest accuracy in benefits assessment is needed. The investment in different techniques may be quite different if the key impact of a regulation is on innovation or on aesthetic degradation.

Knowing the accuracy needed is also critical to determining how much analysis to order. At any point in time it is possible to set some upper and lower limits on how much benefit might be associated with a particular consequence. If both of these extreme values lead to the same policy choice, then no further analysis is needed. If the decision is sensitive to the choice of values, then reducing the range of uncertainty should be the initial objective of analysis. This might be accomplished by a single assessment using a preferred technique,[32] several inexpensive assessments showing the range of values produced by different techniques, or scientific studies improving the accuracy of the consequence estimates upon which all benefits assessments techniques depend.

The initial assessment of extreme values assumes the existence of some previous analyses. From this perspective, every new analysis is part of a chain of analyses, each building upon its predecessors. To improve the long-term efficiency of this enterprise, each assessment should be managed with its successors in mind. Perhaps the best general guidance is to treat benefit assessment as much like a scientific pursuit as possible, with funds being set aside for peer review of all studies,[33] documentation of procedures, and preparation for publication in accessible sources. That peer review should involve diverse individuals, including those skeptical of the assessment technique, and it should be made early in the proceedings as well as at the end (so as to shape the analysis constructively and not just criticize its product).

Basing Decisions on Benefits Assessments

Ultimately, policy making is a gamble, with the policy maker hoping to get the best possible deal for society in a state of incomplete knowledge (Stokey and Zeckhauser 1978; Behn and Vaupel 1983). Benefits assessments aid this process by reducing the uncertainty about facts and formalizing the judgments about values. A policy perspective suggests other contributions as well. One is assessing the magnitude of the uncertainties that remain after all the analyses have been completed, so that policy makers can assess the contingencies for which they must plan.[34] A second is describing how much can ever be known and at what price, enabling policy makers to assess the value of waiting and information gathering. A third is to alert policy makers to the value issues that they must address in order for benefits assessment to proceed, thereby structuring their work. A fourth is detecting biases in the risk analyses upon which benefits assessments depend, such as the tendency for scientists to give "conservative estimates." A fifth is to utilize economics and decision theory to help policy makers avoid common mistakes, such as being trapped by dysfunctional commitment to sunk costs (Teger 1979; Thaler 1980) or the belief that one can talk about "acceptable risks" in the abstract, without consideration of the other consequences associated with the actions producing the risks (Derby and Keeney 1981; Fischhoff 1984).

Decision Research
Arthur D. Little, Inc.

Notes

1. A more detailed scheme for ensuring specification of health and safety regulations may be found in Fischhoff (1984); in application to the specification of one component of such regulations, the term "risk" may be found in Fischhoff, Watson, and Hope (1984).

2. Further discussion may be found in Fischhoff *et al.* (1981) and references therein.

3. Fischhoff (1983) applies this scheme to reveal the residual ambiguities in the Nuclear Regulatory Commission's (1983) recent ambitious attempt to develop "safety goals for nuclear power."

4. As a case in point, consider the problem of controlling the risks from railroad transportation of hazardous materials. One might regulate container design and integrity, safety devices, inspection requirements, train length, track design, and so on. No regulator could hope to identify the best option (or even set of options) from the myriad of possibilities.

5. Crouch and Wilson (1982) provide an insightful discussion of these problems.

4. A Framework for Regulatory Benefits Assessment

6. Harris (1978) describes the regulation of mercury effluents in these terms.

7. These questions also become quite complex, if not intractable, when the affected individuals have different attitudes toward time (i.e. different marginal time preference rates; Sugden and Williams 1978).

8. Popper (1961) makes this observation the core of his philosophy of the social sciences.

9. Of course, if the "compensating risk" leads to increased benefits for the user (e.g., when more stable tractors are used to plow steeper slopes), then it may be a highly rational response (even if it is not the one intended by the regulator). A discussion of the evidence regarding whether such effects are indeed obtained may be found in *Risk Analysis* (1983) No. 2.

10. Discussions of what constitute "appropriate levels of concern" may be found in Fischhoff, Watson, and Hope (1984) and Fischhoff, Svenson, and Slovic (in press).

11. See *Risk Assessment and Risk Assessment Methods* (Merkhofer, Covello, and Menkes 1984).

12. Additional discussion of these issues can be found in Gamble (1978), Hammond and Adelman (1976), Levine (1974), Markovic (1970), and Mazur, Marino, and Becker (1979).

13. Three perspectives are found in Henrion and Fischhoff (in press), Murphy and Winkler (1984), and Wallsten and Budescu (1983).

14. Exactly how this is done is itself a complex topic around which a substantial technical literature has grown. One issue, the question of whom to include at all in households and firms, has been discussed above in the section on identifying relevant consequences.

15. The total asset position is determined by the individual's income, endowment of owned goods, and opportunity to find substitutes and complements of the good in question.

16. Discussion of the empirical and ethical issues surrounding this topic can be found in Knetsch (1984) and Knetsch and Sinden (in press).

17. One reason for objecting to the Pareto criterion would be the feeling that the venture should offer something to the nonlosers, particularly in those situations where the winners have more power to create options. For empirical evidence regarding the acceptability of this principle, see McClelland and Rohrbaugh (1978).

18. These issues arise even when the potential Pareto improvement criterion is restricted to transferable commodities (such as money). Chapter 8 discuses how they emerge with measures such as consumer surplus, compensating variation, and equivalent variation, which rely on the information in market demand curves. Chapter 6 examines willingness-to-pay and willingness-to-accept measures for nonmarket, public goods. A review of the ethical issues involved in payment for work-place hazards may be found in Derr *et al.* (1983).

19. Some writers use the term *value* to describe such preferences in the case of known outcomes and *utility* for preferences over uncertain outcomes. We will use *utility* to cover both cases.

20. A good general introduction is Keeney and Raiffa (1976). A worked-out example applying to risk is found in Fischhoff, Watson, and Hope (1984). A detailed adaption of the methodology to decisions regarding the siting of energy facilities is the topic of Keeney (1980).

21. The procedures cannot eliminate the influence of current conditions insofar as the poor will be able to focus on other elements when their financial security is assured.

22. A cacaphonic acronym for this condition is UIIGI, for "unanimous indifference implies group indifference."

23. Fuller expositions of empirical tests of these ambitious assumptions can be found in Fischhoff, Goitein, and Shapira (1982), Schoemaker (1984), Simon (1978), Slovic, Fischhoff, and Lichtenstein (in press).

24. See Fischhoff, Slovic, and Lichtenstein (1980, 1981), Kahneman, Slovic, and Tversky (1982), Tversky and Kahneman (1974).

25. See, e.g., Fischhoff, Slovic, and Lichtenstein (1980), Hogarth (1982), Rokeach (1973), Tversky and Kahneman (1981), National Research Council (1982), Turner, Martin, and De Maio (1985), Hershey and Schoemaker (1980), Grether and Plott (1979), Slovic and Lichtenstein (1983).

26. Beach and Mitchell (1978) and Payne (1982) offer schemes for predicting which decision process people will pick from their repertoire of possibilities.

27. On public opinion polls see Conn (1983) and especially the papers by Farhar-Pilgrim and Melber therein.

28. In response, a thoughtful interviewee might ask, "What alternatives should I be considering? Am I allowed to worry about equity issues? Do you mean the current technology or what it might be if the responsible critics were taken seriously?"

29. See n. 20 above.

30. Especially important is the degree of multicollinearity, or correlation among predictor variables (e.g., the tendency for good features of jobs to go together). When this happens, the respective roles of the correlated variables are very hard to discern.

31. Further discussion of these approaches may be found in Chapter 5 of Fischhoff *et al* (1981).

32. The techniques of value-of-information analysis can be used to formalize this relationship (Raiffa 1968).

33. A similar suggestion regarding another applied science – survey research – can be found in National Research Council (1982).

34. Uncertainty assessment is a major topic in the companion volume, Merkhofer, Covello, and Menkes (1984). Insofar as benefits assessment is a scientific pursuit, the issues that it faces are the same as those encountered by the risk analyses upon which it depends.

References

Ajzen, I. and M. Fishbein. 1977 'Attitude-Behavior Relations: A Theoretical Analysis and Review of Empirical Research.' *Psychological Bulletin* 84: 888–918.
Beach, L. R. and T. R. Mitchell. 1978. 'A Contingency Model for the Selection of Decision Strategies.' *Academy of Management Review* 3: 439–449.
Behn, R. D. and J. W. Vaupel. 1983. *Quick Analysis for Busy Decision Makers*. New York: Basic Books.

Bell, D. 1982. 'Regret in Decision Making under Uncertainty.' *Operations Research* 30: 961–981.
Bradburn, N. M. and S. Sudman, 1979. *Improving Interview Method and Questionnaire Design.* San Francisco: Jossey-Bass.
Brookshire, D. S., B. C. Ives, and W. D. Schulze. 1976. 'The Valuation of Aesthetic Preferences.' *Journal of Environmental Economics and Management* 3: 325–346.
Brown, R. V. 1975. *Modeling Subsequent Acts for Decision Analysis.* DDI Technical Report 75–1. McLean, Va.: Decisions and Designs, Inc.
Cohen, B. and W. Lee. 1979. 'A Catalog of Risks.' *Health Physics* 36: 707–722.
Cohen, J. and P. Cohen. 1975. *Applied Multiple Regression/Correlation Analysis for the Behavioral Sciences.* Hillsdale, N.J.: Erlbaum.
Conn, W. D., ed. 1983. *Conserving Energy and Material Resources.* Boulder: Westview.
Coombs, C. H. 1964. *A Theory of Data.* New York: Wiley.
Crouch, E. A. C. and R. Wilson. 1982. *Risk/Benefit Analysis.* Cambridge, Mass.: Ballinger.
Dawes, R. M. 1979. 'The Robust Beauty of Improper Linear Models in Decision Making.' *American Psychologist* 34: 571–582.
———, J. Delay, and W. Chaplin. 1976. 'The Decision to Pollute.' *Environment and Planning* 6: 3–10.
Derby, S. L. and R. L. Keeney. 1981. 'Risk Analysis: Understanding "How Safe is Safe Enough?"' *' Risk Analysis* 1: 217–224.
Derr, P., R. Goble, R. E. Kasperson, and R. W. Kates. 1983. 'Responding to the Double Standard of Worker/Public Protection.' *Environment* 25(6): 35–36.
Einhorn, H. J. and R. M. Hogarth. 1978. 'Confidence in Judgment: Persistence of the Illusion of Validity.' *Psychological Review* 86: 465–416.
Elliott, G. R. and C. Eisdorfer. 1982. *Stress and Human Health.* New York: Springer.
Ericsson, A. and H. Simon. 1980a. 'Verbal Reports as Data.' *Psychological Review* 87: 215–251.
———, ———. 1980b. *Verbal Reports as Data.* New York: Academic Press.
Evans, N. and C. W. Hope 1983. 'Costs for Nuclear Accidents: Implications for Reactor Choice.' *Energy Policy* 11: 215–304.
Feyerabend, P. 1975. *Against Method.* London: Verso.
Fischer, G. W. 1977. 'Convergent Validation of Decomposed Multiattribute Utility Assessment Procedures for Risky and Riskless Decisions.' *Organizational Behavior and Human Performance* 18: 295–315.
Fischhoff, B. 1975. 'Hindsight ≠ Foresight: The Effect of Outcome Knowledge on Judgment under Uncertainty.' *Journal of Experimental Psychology: Human Perception and Performance* 1: 288–299.
———. 1980a. 'Clinical Decision Analysis.' *Operations Research* 28: 28–43.
———. 1980b. 'For Those Condemned to Study the Past: Reflections on Historical Judgment.' In R. A. Shweder and D. W. Fiske, eds., *New Directions for Methodology of Behavior Science: Fallible Judgment in Behavioral Research.* San Francisco: Jossey-Bass.
———. 1982. 'Debiasing.' In D. Kahneman, P. Slovic, A. Tversky, eds., *Judgment under Uncertainty: Heuristics and Biases.* New York: Cambridge University Press.
———. 1983. 'Safety Goals for Nuclear Power.' *Journal of Policy Analysis and Management*: 559–575.
———. 1984. 'Setting Standards: A Systematic Approach to Managing Public Health and Safety Risks.' *Management Science* 30: 823–843.
———, B. Goitein, and A. Shapira. 1982. 'The Experienced Utility of Expected Utility Approaches.' In N. Feather, ed., *Expectancy, Incentive and Action.* Hillsdale, N. J.: Erlbaum.
———, S. Lichtenstein, P. Slovic, S. Derby, and R. Keeney. 1981. *Acceptable Risk.* New York: Cambridge University Press.
———, P. Slovic, and S. Lichtenstein. 1980. 'Knowing What You Want: Measuring Labile Values.' In T. Wallsten, ed., *Cognitive Processes in Choice and Decision Behavior.* Hillsdale, N.J.: Erlbaum.

———, ———, ———. 1981. 'Lay Foibles and Expert Fables in Judgments about Risk.' *American Statistician* 36: 240–255.
———, O. Svenson, and P. Slovic. In press. 'Active Responses to Environmental Hazards.' In D. Stokols and I. Altman, eds., *Handbook of Environmental Psychology*. New York: Wiley.
———, S. Watson, and C. Hope. 1984. 'Defining Risk.' *Policy Sciences* 17: 123–139.
Fishburn, P. C. 1982. *The Foundations of Expected Utility Theory*. Dordrecht: D. Reidel.
Fox, J. 1980. 'Making Decisions under Influence of Memory.' *Psychological Review* 87: 190–211.
Gamble, D. J. 1978. 'The Berger Inquiry: An Impact Assessment Process.' *Science* 199: 946–951.
Gibson, M., ed. In press. *Risk, Consent, and Air*. Totowa, N.J.: Rowman & Allenheld.
Glass, G. V., B. McGaw, and M. L. Smith. 1981. *Metaanalysis in Social Research*. New York: Sage.
Goldberg, L. R. 1968. 'Simple Models or Simple Processes? Some Research on Clinical Judgments.' *American Psychologist* 23: 483–496.
Goodin, R. E. 1978. 'Uncertainty as an Excuse for Cheating our Children.' *Policy Sciences* 10: 25–48.
Grether, D. M. and C. R. Plott. 1979. 'Economic Theory of Choice and the Preference Reversal Phenomenon.' *American Economic Review* 69: 623–638.
Hammond, K. R. and L. Adelman. 1976. 'Science, Values and Human Judgment.' *Science* 194: 389-396.
———, and D. A. Summers. 1972. 'Cognitive Control.' *Psychological Review* 79: 58–67.
Hardin, G. 1968. 'The Tragedy of the Commons.' *Science* 162: 1243–1248.
Harris, S. R. 1978. 'Mercury: Measuring and Managing the Hazard.' *Environment* 20 (8).
Hedges, L. V. and V. Olkin. 1980. 'Vote-Counting Methods in Research Synthesis.' *Psychological Bulletin* 88: 359–369.
Henrion, M. and B. Fischhoff. In press. 'Uncertainty Assessment in the Estimation of Physical Constants'. *American Journal of Physics*.
Hershey, J. C. and P. J. H. Shoemaker. 1980. 'Risk Taking and Problem Context in the Domain of Losses: An Expected Utility Analysis.' *Journal of Risk and Isurance* 47: 111–132.
Hogarth, R. ed. 1982. *New Directions for Methodology of Social and Baviral Science: The framing of questions and the Consistency of Response*. San Francisco: Jossey-Bass.
Janis, I. L., ed. 1982. *Counseling on Personal Decisions*. New Haven: Yale University Press.
Kahneman, D. and A. Tversky. 1979. 'Prospect Theory.' *Econometrica*.
———, P. Slovic, and A. Tversky. 1982. *Judgment under Uncertainty*: *Heuristics and Biases*. New York: Cambridge University Press.
Kates, R. W. 1978. *Risk Assessment of Environmental Hazard*. Chichester, England: Wiley.
Keeney, R. L. 1980. *Siting Energy Facilities*. New York: Academic Press.
——— and H. Raiffa. 1976. *Decisions with Multiple Objectives: Preferences and Value Tradeoffs*. New York: Wiley.
Knetsch, J. L. 1984. 'Legal Rules and the Basis for Evaluating Economic Losses.' *International Review of Law and Economics* 4: 5–13.
——— and J. A. Sinden. In press. 'Willingness to Pay and Compensation Demanded.' *Quarterly Journal of Economics*.
Krantz, D. H., R. D. Luce, P. Suppes, and A. Tversky. 1971. *Foundations of Measurement*, Vol. I. New York: Academic Press.
Krutilla, J. V. and A. C. Fisher. 1975. *The Economics of Natural Environments*. Baltimore: Johns Hopkins University Press.
Kunreuther, H., R. Ginsberg, L. Miller, P. Sagi, P. Slovic, B. Borkin, and N. Katz. 1978. *Disaster Insurance Protection*: *Public Policy Lessons*. New York: Wiley.
Levine, M. 1974. 'Scientific Method and the Adversary Model: Some Preliminary Thoughts.' *American Psychologist* 29: 61–716.
Lichtenstein, S., B. Fischhoff, and L. D. Phillips. 1982. 'Calibration of Probabilities: The State of the Art to 1980.' In D. Kahneman, P. Slovic, and A. Tversky, eds., *Judgement under*

Uncertainty: Heuristics and Biases. New York: Cambridge University Press.
Lind, R. C., K. J. Arrow, G. Corey, P. Dasgupta, A. Sen, T. Stauffer, J. E. Stiglitz, J. Stockfish, and R. Wilson. 1982. *Discounting for Time and Risk in Energy Policy.* Baltimore: Johns Hopkins University Press.
Mahoney, M. J. 1979. 'Psychology of the Scientist: An Evaluative Review.' *Social Studies of Science.* 9; 349–375.
Markovic, M. 1970. 'Social Determinism and Freedom.' In H. E. Kelfer and M. K. Munitz, eds., *Mind, Science and History.* Albany: State University of New York Press.
Mazur, A. 'Disputes between Experts.' *Minerva* 11: 243–262.
———, A. A. Marino, and R. O. Becker. 1979. 'Separating Factual Disputes from Value Disputes in Controversies over Technology.' *Technology in Society* 1: 229–237.
McClelland, G. H. and J. Rohrbaugh. 1978. 'Who Accepts the Pareto Axiom?' *Behavioral Science* 23: 446–456.
McKean, R. N. 1972. 'The Use of Shadow Prices.' In R. Layard, ed., *Cost-Benefit Analysis.* Harmondsworth: Penguin.
Meehl, P. E. 1954. *Clinical vs. Statistical Prediction.* Minneapolis: University of Minnesota Press.
Merkhofer, M., V. Covello, and J. Menkes, eds. 1984. *Risk Assessment and Risk Assessment Methods.* Report prepared for the National Science Foundation. Boston: Charles River Associates.
Miller, G. A. 1956. 'The Magical Number Seven, Plus or Minus Two: Some Limits on Our Capacity for Processing Information.' *Psychological Review* 63: 81–97.
Mishan, E. J. 1976. *Cost-Benefit Analysis.* New York: Praeger.
Murphy, A. H. and R. L. Winkler, In press. 'Probability of Precipitation Forecasts.' *Journal of the American Statistical Association.*
National Research Council. 1982. *Survey Measure of Subjective Phenomena.* Washington, D.C.
———. 1983. *Priority Mechanisms for Toxic Chemicals.* Washington, D.C.
Nelkin, D., ed. 1984. *Controversy: Politics of Technical Decisions.* Beverly Hills: Sage.
Newell, A. and H. A. Simon. 1972. *Human Problem Solving.* Englewood Cliffs, NJ: Prentice-Hall.
Nisbett, R. W. and T. D. Wilson. 1977. 'Telling More than We Can Know: Verbal Reports on Mental Processes.' *Psychological Review* 84: 231–259.
Nuclear Regulatory Commission. 1983. *Safety Goals for Nuclear Power Plants: A Discussion Paper.* NUREG–0880. Washington, D.C.
Otway, H. J. and J. J. Cohen. 1975. *Revealed Preferences: Comments on the Starr Benefit-Risk Relationships.* Research Memorandum 75–5. Laxenburg, Austria:International Institute for Applied Systems Analysis.
Page, R. T. 1978. 'A Generic View of Toxic Chemicals and Similar Risks.' *Ecology Law Quarterly* 7: 207–243.
———. 1981. 'A Framework for Unreasonable Risk in the Toxic Substances Control Act.' In R. Nicholson, ed., *Carcinogenic Risk Assessment.* New York: New York Academy of Sciences.
Payne, J. W. 1982. 'Contingent Decision Behavior.' *Psychological Bulletin* 92: 382–401.
Payne, S. L. 1952. *The Art of Asking Questions.* Princeton: Princeton University Press.
Pigou, L. 1932. *The Economics of Welfare.* London: Macmillan.
Popper, K. R. 1961. *The Logic of Scientific Discovery.* New York: Science Editions.
Raiffa, H. 1968. *Decision Analysis.* Reading. Mass.: Addison-Wesley.
Rokeach, M. 1973. *The Nature of Human Values.* New York: Free Press.
Schelling, T. C. 1980. *Micromotives and Macrobehavior.* New York: Norton.
Schulze, W. 1974. 'Social Welfare Functions for the Future.' *American Economist* 18(1): 70–81.
Schuman, H. and M. Johnson. 1976. 'Attitudes and Behavior.' *Annual Review of Sociology* 2: 161–207.
Simon, H. A. 1956. 'Rational Choice and the Structure of the Environment.' *Psychological*

Review 63: 129–138.

———. 1978. 'Rationality as Process and Product of Thought.' *American Economic Review* 68: 1–16.

Slovic, P. and S. Lichtenstein. 1983. 'Preference Reversals: A Broader Perspective.' *American Economic Review* 73: 596–605.

———, B. Fischhoff, and S. Lichtenstein. 1979. 'Rating the Risks." *Environment* 213: 14–20, 36–39.

———, ———, ———. 1980. 'Facts and Fears: Understanding Perceived Risk.' In R. Schwing and W. A. Albers, Jr., eds., *Societal Risk Assessment: How Safe is Safe Enough?* New York: Plenum.

———, ———, ———. 1984. 'Modeling the Societal Impact of Fatal Accidents.' *Management Science* 30: 464–474.

———, ———, ———. In press. 'Decision Making.' In R. C. Atkinson, R. J. Herrnstein, G. Lindzey, and R. D. Luce, eds., *Stevens' Handbook of Experimental Psychology*. 2nd ed. New York: Wiley.

Starr, C. 1969. 'Social Benefit Versus Technological Risk.' *Science* 165: 1232–1238.

Stokey, E. and R. Zeckhauser. 1978. *A Primer of Policy Analysis*. New York: Norton.

Sugden, R. and A. Williams. 1978. *The Principles of Practical Cost Benefit Analysis*. Oxford: Oxford University Press.

Svenson, O. 1979. 'Process Descriptions of Decision Making.' *Organizational Behavior and Human Performance* 23: 86–112.

Teger, A. I. 1979. *Too Much Invested to Quit*. New York: Pergamon.

Thaler, R. H. 1980. 'Toward a Positive Theory of Consumer Choice.' *Journal of Economic Behavior and Organization* 1: 38–60.

——— and H. M. Shefrin. 1981. 'An Economic Theory of Self Control.' *Journal of Political Economy* 89: 392–406.

Turner, C., E. Martin, and T. De Maio, eds. 1985. *Survey Measure of Subjective Phenomena*. New York: Sage.

Tversky, A. 1972. 'Elimination by Aspects: A Theory of Choice.' *Psychological Review* 79: 281–299.

——— and D. Kahneman. 1974. 'Judgment under Uncertainty: Heuristics and Biases.' *Science* 211: 453–458.

———, ———. 1981. 'The Framing of Decisions and the Psychology of Choice.' *Science* 211: 453–458.

Von Winterfeldt, D. and W. Edwards. 1982. 'Costs and Payoffs in Perceptual Research.' *Psychological Bulletin* 91: 609–622.

Wallsten, T. and D. Budescu. 1983. 'Encoding Subjective Probabilities: A Psychological and Psychometric Review.' *Management Science* 29: 151–173.

Wilson, R. 1979. 'Analyzing the Daily Risks of Life.' *Technology Review* 81(4): 40–46.

Chapter 5

Theory of Regulatory Benefits Assessment: Econometric and Expressed Preference Approaches

Louis Anthony Cox, Jr.

The preceding Chapter 4 has outlined a general framework and a sequence of steps for conducting a benefits assessment. These are summarized below together with several important policy questions that must be confronted in any benefits assessment for a major regulation:

1. *Define the scope of the assessment.* What benefits (what effects of what regulation on which stakeholders) are to be assessed? What empirical assumptions and technical specifications are to be made?

2. *Identify and describe policy alternatives* to the option being assessed. What attributes should be used in describing and assessing the potential consequences of each alternative?

3. *For each alternative, identify the probable resulting welfare changes for each stakeholder group.* Should benefits to each stakeholder group be defined subjectively in terms of the preferences and perceptions of its members, or objectively in terms of measurable changes in physical and economic variables?

4. *Aggregate the probable welfare consequences to each stakeholder group* to obtain a measure of the net social benefit from each alternative. How should benefits to each group be aggregated over time and in the presence of uncertainties? What risk attitude should be taken?

It should be noted that this framework is suitable when the set of policy alternatives is small. When there are many proposed alternatives, it is no longer practical to assess the benefits of each in detail, and more efficient decision procedures based on dominance and sequential eliminations must be used. The structuring of an efficient decision procedure involves questions of cost estimation and optimal information collection that go well beyond the problems of benefits assessment per se.

In this chapter and the next, we review methods that have been developed for accomplishing the steps outlined above and that help to answer some of the policy questions if certain technical specifications are accepted. We shall assume that appropriate modeling and description of the probable objective consequences of a regulation has already been done (see, e.g., Merkhofer, Covello, and Menkes 1984), although uncertainties may remain. We are concerned with moving from these descriptions to the resulting *evaluations* of the corresponding changes in benefits, or social welfare.

Both this chapter and Chapter 6 focus on assessment of benefits for stakeholder groups or individuals, and aggregation of the resulting benefit estimates across groups or individuals. This chapter deals with concepts and methods that cut across different formal and standardized approaches or specific techniques of benefits assessment; Chapter 6 considers methods that integrate these ideas into well-defined approaches that currently appear to be of most use to practitioners. Our hope is that the state-of-the-art concepts discussed in this chapter will be used to apply or extend the approaches in Chapter 6 in new and challenging ways. This reflects our belief that benefits assessment is still very much an evolving art. Figure 5–1 (on page 107) provides an overview of the principal approaches considered.

Benefits assessments of regulatory programs and of specific regulations are generally undertaken either *retrospectively*, for the purpose of estimating the productivity of different programs in increasing social welfare, so as to provide information to assist in the allocation of resources to agencies and programs; or *prospectively*, for the purpose of obtaining information that will help in deciding what future regulatory actions to take. Retrospective assessments are useful in evaluating the performance of regulatory agencies and decision-making *processes*; while prospective assessments are more appropriate for evaluating specific regulatory *proposals*. In the context of Executive Order 12291, prospective actions ("major regulations") are of primary concern. We will therefore focus on the design and implementation of benefits assessments that are intended to guide choice among alternative proposed regulatory actions.

Use of Individual Preferences to Define Benefits

Most of the benefits-assessment methods that we shall be reviewing make the fundamental assumption (or technical specification) that *individual preferences are to be used as the basis for evaluating the benefits from a regulation*. This is defensible if the following conditions exist:

1. Informed rationality in individual behavior.
2. Harmony between individual and collective rationality (e.g., no externalities).
3. Truthful revelation of preferences (e.g., no strategic misrepresentation of preferences for the purpose of manipulating the social decision).
4. Consumers offered a set of choices rich enough for their responses to reveal relevant preferences rather than just constraints.

5. Theory of Regulatory Benefits Assessment

In practice, as discussed in Chapter 4, these conditions are often not fully met, and other frameworks for social decision making may be useful. Other possible specifications on which assessment methods are based include:

1. Using the preferences of one or a few public decision makers who represent the public and act in the public interest; this is sometimes called the supra-decision-maker approach (Keeney and Raiffa 1976).
2. Using the social preferences that are presumably revealed by past policy choices.
3. Using the collective preferences defined by the results of social decision processes such as voting, arbitrated negotiation and settlement, or tort-law adjudication.

Although these approaches fall within the broad scope of political economy, they are outside the purview of the economic cost-benefit framework described in this book. It should be noted, however, that these alternative frameworks may be as valuable as the cost-benefit framework for guiding public decisions, especially those which may have a paternalistic rationale (such as the regulations of certain foods and drugs).

From another perspective, we can ask of a social decision process or institution whether each individual expects to gain on average from its use even though he may lose on some occasions. In this approach, social decisions are not evaluated in isolation, based on the costs and benefits of each alone. Rather, each decision is seen as part of a sequence of decisions generated by a socially beneficial decision process or mechanism, possibly involving implicit compromises in which those who lose on one round win on the next, and everyone wins on average and in the long run. Fairness and efficiency are evaluated for the sequence, rather than for each decision in it. Although this *process approach* to benefits assessment provides a fruitful and useful perspective, it generally falls outside the scope of this discussion. We focus on the assessment of individual regulations, rather than of regulatory processes.

Expressed Rather than Revealed Preferences

As discussed in Chapter 4, benefits assessment for regulations that supply public or other nonmarket goods – such as many antipollution or safety regulations – requires that individual preferences be either inferred from observed market behavior or elicited by questioning. In general, however, it is difficult to infer preferences for nonmarket goods from preferences for market goods, except under very strong assumptions about individual preferences. With few exceptions, benefits assessment methods for public goods rely on surveys, interviews, petitions, and other devices through which preferences are actively expressed or announced, rather than passively revealed.

Developing Technical Specifications for Representation of Individual Preferences

If benefits are to be defined in terms of individual preferences, then technical specifications are required to determine the following:

1. Which individuals are to have their preferences represented.
2. How these preferences are to be turned into measures of social benefit.
3. What units are to be used to express both preferences and benefits, assuming that they are measured in the same units.
4. How preferences of different qualities are to be treated.

For example, should the sharply defined preferences of well-informed individuals be treated the same way as the vague and uncertain preferences of poorly informed people or of future generations? How much information about relevant issues must an individual have (and hence, perhaps, must the eliciter of preferences provide) before his resulting preferences become a useful guide for policy making?

There are no objective methods for making these four specifications. However, technical analysis provides some limited guidance. Question 1 is elaborated upon in the section that follows. Two alternative approaches to resolving questions 2 and 3 – the econometric approach and the social utility approach – are considered in other sections. Also reviewed later is multi-attribute assessment, which uses aspects of both the econometric and the social utility approach to deal with question 4.

Selection of a sample population

Benefits-assessment methods that rely on announced rather than revealed preferences require selection of a sample of respondents. The selection must respect three concerns: political concerns about the representativeness of the preferences in the sample group; statistical concerns about whether the sample is large enough to protect against erroneous conclusions; and practical concerns about whether the sample can be probed with reasonable expenditures of resources.

Adequate political representation

The sample selected must adequately represent the stakeholder groups whose interests and preferences are to be considered. A key principle of cost-benefit analysis is that the interests of *all* affected stakeholders be taken into account in making social decisions. This is sometimes described by saying that all of the benefits (and costs) to all of the individuals, both past and present, affected by a regulation must be accounted for (Stokey and Zeckhauser 1978). Such a requirement is clearly impossible to meet, but it emphasizes the importance of considering all identifiable interests in a fair and impartial way.

Unfortunately, technical analysis provides little guidance on how to construct an adequately representative sample of respondents, especially since the interests and preferences of future generations are necessarily speculative. Some theoretical results on the minimum number of people required to represent a given number of partially overlapping stakeholder groups (i.e., results on how to construct a scaled-down sample that represents the set of overlapping interests in microcosm) are available but seem to be of little direct practical value (March 1966; Banzhaf 1968). From a pragmatic perspective, the sample should be designed so that no stakeholder group that stands to lose from the contemplated regulation is excluded.

Another consideration in designing politically responsive samples has to do with the amount of diversity or *conflict* in the population being sampled. Benefit-cost analysis is, among other things, useful as a device for conflict clarification and resolution. If all parties feel that their interests have been fairly and fully represented in the calculation of net social benefits, then the decisions implied or supported by a benefits assessment are more likely to be politically acceptable. Additionally, the process of benefits assessment may itself contribute to reduce or increase conflict, depending on how it is conducted. For example, asking each respondent whether he is for or against a specific proposed regulation is apt to uncover less decision-irrelevant conflict than asking each respondent to identify the best or most-preferred *level* of a public good. On the other hand, if the benefits assessment is designed to uncover information that can be used to refine the proposed regulation, then the latter may be more useful. If the assessment can be designed to minimize the conflict engendered, then the sample size required to adequately represent parties with partially conflicting views can also be reduced (Abonyi 1983).

Statistical adequacy of the sample

A second set of related concerns deals with sample size from the standpoint of statistical representativeness. That is, it is important to assure with a reasonable degree of confidence that the results obtained from the selected sample are correct, in the sense that they give the same results that would have been obtained if preferences had been elicited from the whole population.

The basic problem is the same in any statistical sampling situation: the distribution of preferences in the sample may, by the luck of the draw, differ enough from the distribution of preferences in the whole population so that conclusions about changes in welfare based on the sample are incorrect for the whole population. The usual defense against this sort of error is a proper research design which minimizes the expected error of estimation. What makes sample selection difficult in practice is that the population affected by a regulation is never homogeneous. Not only do different stakeholder groups have different preference distributions, so that simple sampling schemes must be revised to make sure that the full mix is captured, but the stakeholder groups and their sizes may be difficult to identify.

Tracing the economic effects of a regulation on affected *groups* may be easy. For example, the groups can often be identified by tracing the input-output flows connecting the directly affected industries or economic agents to the rest of the economy, and then accounting for externalities at the end. Ideally, however, we would like to count *individuals*, which may be difficult. Each affected individual should be counted exactly once. But individuals may belong to several stakeholder groups. Thus, it would be a mistake to first examine the health effects of a proposed occupational safety regulation on a firm's employees and then separately its financial effects on the firm's stockholders if the stockholders were the employees. The integrated preferences of the individuals involved are what counts. Such double counting can only be avoided if overlaps in the memberships of different economically defined stakeholder groups affected by a regulation can be estimated or safely ignored.

Stratified sampling techniques are in principle available to overcome the problem of inadequate representation of multiple stakeholder groups, especially if the decision to be made depends only on the *average* of individual preferences (if preferences are expressed numerically) or on the total number of people favoring one alternative over another (if preferences are expressed through pairwise ordinal comparisons). However, these techniques are apt to be of little use unless the stakeholder groups to be sampled can be clearly identified before the sample has been taken (Cochran 1953).

Inadequate stratification can lead to underrepresentation of small interest groups and to paradoxes of aggregation. For example, aggregate data could show that nonsmokers as a group, in aggregate, favor a particular health or safety regulation more than smokers. Yet it might nonetheless be the case that in every particular stakeholder group – as specified by a complete set of attributes such as age, health status, income level, and number of years as a smoker – the smokers favor the regulation more than the nonsmokers. This phenomenon, known in statistics as Simpson's Paradox, can arise if we define the strength with which the group favors a proposed regulation as the proportion of the group that favors it. Similar paradoxes of aggregation arise under other measures of group preference if the sample is inadequately stratified.

Practical concerns in sample selection
In many assessments the problems of proper stratification and adequate sample size are largely academic: practical constraints on who is willing and available to participate in the sample are more important in determining the final sample size and composition. For example, people with a vested interest in the outcome are more apt to participate in petitions, voluntary surveys, or other data collection efforts than are people with no direct stake.

5. Theory of Regulatory Benefits Assessment

Such self-selection can threaten the representativeness of the sample, making reliable inferences about the distribution of preferences in the overall population difficult. On the other hand, conducting a rigorously designed sample survey may not be possible given that participation is voluntary and that the resources available for sampling of individual preferences are apt to be very limited.

Summary of sample design criteria

The design of a sample for assessing relevant preferences must reflect an awareness of

1. Diversities and nonhomogeneities in the population being sampled. The sample must be large enough and sufficiently well constructed to adequately represent the conflicting interests involved.
2. Sample sizes must be large enough to provide stable and reliable estimates of stakeholder preference distributions.
3. Affected individuals must not be counted more (or less) than once. An individual who is affected by a regulation in several capacities must not be accidently counted several times in the process of identifying affected parties.
4. Self-selection and similar biases in sample design should be avoided or controlled (Campbell and Stanley 1963).

Despite these difficulties, for some health, safety, and environmental regulations the affected stakeholder groups and individuals are quite easily identified. Two examples follow.

Example 1: Identifying the target population for an OSHA regulation

The population affected by an Occupational Safety and Health Agency (OSHA) regulation is often quite specific and easily identified as the workers in one or a few industries or industrial processes. If we accept the policy that only the directly affected workers should be consulted about the value of the health or safety benefits provided by a proposed regulation, then the population to be sampled can be sharply and conveniently defined. The question of whether the preferences of other groups having no direct safety stake (or perhaps no direct economic stake) in the outcome should also be incorporated into OSHA's decision-making process is a technical specification for the benefits assessment. It requires a policy decision. For example, using a "value of life" – or, more accurately, a value for a small reduction in the risk of death – calculated from preferences revealed in a broader population and for different types of risk may be judged inappropriate and possibly irrelevant for the particular worker population and type of risk involved. On the other hand, if the specific worker population involved places a higher value on small risk reductions than do other members of society, then it is not clear that the evaluations of the affected group alone should be used as the sole basis for a benefits assessment that will be used to guide the allocation of society's resources.

At issue here is the policy question of who is to be considered to have a legitimate stake in the outcome of a regulation and hence to be entitled to have his preferences represented in the benefits assessment. Are legitimate stakeholders only those who are directly affected by a regulation, or are taxpayers who help to provide the resources necessary for regulation thereby entitled to inject their own preferences and attitudes toward the directly affected parties into the assessment? The question of who is to be (or should be) represented in the benefits assessment is essentially political and must be clearly specified in the assessment.

Example 2: Implicit identification of the target population for a CPSC regulation

Many regulations implictly identify a target population. For example, the Lawn Mower Safety Standard of the Consumer Product Safety Commission (CPSC) is implicitly addressed to the population of current and potential lawnmower owners and operators. For quantitative benefits assessment, however, it is necessary to estimate the size of such implicitly defined populations so that the benefits estimated through sampling can be scaled up appropriately. Once again, these questions remain: (1) how finely to partition the target population into strata based on other relevant factors; (2) which other populations to include (e.g., potential lawnmower purchasers who dropped out of the market because of increased prices) in evaluating the worth of reduced injuries to lawnmower operators. Statistical theory can help answer the first question but not the second. However, in this case, the product in question is a market good, and market data on quantity demanded, extent of lost sales, and so forth are available to support traditional cost-benefit analyses (Miller and Yandle 1979: Chapter 5).

The Econometric Approach

Given the practical and policy problems associated with construction or selection of an adequate sample for assessing preferences, it is perhaps not surprising that many agency benefits assessments have avoided the sampling question altogether. Rather than literally specifying the set of legitimate stakeholders and collecting preference data from them, most assessments employ an indirect strategy in which

1. The probable physical and economic effects of the proposed regulation are forecast from models or rough calculations. These effects might include the expected reduction in the average numbers of accidents per year by severity class or the expected number of man-years saved by measures that improve public health.

5. Theory of Regulatory Benefits Assessment

2. The values of these projected changes in attribute levels are assessed "objectively," rather than on the basis of survey questions. For example, the range of dollar values of a statistical life saved, adjusted for age and health status, may be estimated on the basis of past studies of revealed aggregate wage-risk tradeoffs obtained from market data, or on the basis of announced individual willingness-to-pay data from other studies.

3. The predicted changes in risk levels are weighted by the values of these changes, as estimated from past studies or as imputed from market data for similar cases. These estimated values are used as surrogates for the prices that would be supplied by the market if the benefits of the regulation could be sold as market goods.

4. The assessed benefit of the proposed regulation is then found by analogy with market pricing as the sum of the values of the changes in each attribute, weighted by the surrogate prices or values imputed to each. Typically, high and low values from several past studies are used as weights, so that an interval estimate of benefits is the final result.

For ease of reference, we will call this strategy for assessing the benefits of health, safety, and environmental regulations the *econometric approach*. Its distinguishing features are that

1. New preference data for the specific stakeholder groups affected and the specific regulation being evaluated are not collected. The values or surrogate prices estimated in previous studies are assumed to be transferable across contexts.

2. Uncertainties about the effects of a proposed regulation (and about the values to be applied to changes in different attributes) propagate through the analysis and appear at the end as uncertainties about the magnitude of the benefits to be obtained from the regulation.

3. Changes in attribute levels are valued and aggregated additively, by analogy with market prices. If attribute levels are changed from $x_1, \ldots, x_n$ to $x_1^*, \ldots, x_n^*$ by a regulation, where x_i denotes the initial level of the ith attribute, then the resulting benefit is defined as $\Sigma_i v_i x_i^* - \Sigma_i v_i x_i = \Sigma_i v_i (x_i^* - x_i)$, where v_i is the surrogate price estimated for changes in the ith attribute level.

The econometric approach saves the costs of new data collection and is well suited to quick approximate calculations. It gives results that are objective in the sense that they can be checked by reference to past studies and have a natural market interpretation. The econometric strategy is also quite flexible: the ways in which the surrogate prices or value coefficients v_i are initially determined are largely left to the ingenuity of the investigator.

Transfer of Preferences across Contexts

The econometric approach relies on very strong assumptions. The assumption that preferences expressed by one population in one situation can be transferred to another group in a different situation requires careful justification. Specifically, the practitioner should avoid the common fallacy of applying preferences revealed by individuals who have a choice of what risks or disbenefits to accept – for example, workers choosing among occupations or house buyers choosing among locations – to situations where similar risks or disbenefits are imposed by an externality (such as pollution effects on homeowners who are not in the market for new houses) over which the individuals have no control. The value of a reduction in a voluntarily assumed first-party risk – for example from cigarette smoking or driving without a seatbelt – may be very different from the value of a comparable reduction in a risk that is accepted as part of an economic transaction (a second-party risk) or that is imposed via an externality by the actions of others (a third-party risk). The value of a statistical life saved through risk reduction depends on the nature of the risk that has been reduced. At best, personal valuations of small risk reductions should be transferred within, but not between, risk categories.

Treatment of Uncertainty

The econometric strategy for the treatment of uncertainty can also be questioned. The usual reasoning is that if the value of a reduction in accident frequency is $1,000 per accident prevented, and if the expected number of accidents prevented by a proposed regulation is equally likely to be anywhere from 2 to 20 per year, then the expected value of the regulation from accidents prevented is equally likely to be anywhere from $2,000 to $20,000 per year. The difficulty with this reasoning is that for most individuals the value of the uncertain benefits offered by a proposed regulation is different from the expected value of the benefits offered. How different depends on how risk-averse the individuals are, and hence what risk premiums they would require to compensate them for bearing risks. The strategy of assessing benefits for known risk reductions (or known changes in other attribute levels) and then translating uncertainties about the magnitudes of these changes directly into uncertainties about the magnitudes of the corresponding benefits requires the strong assumption that individual preferences are neutral toward the uncertainty itself. An alternative approach would be to elicit individual valuations of uncertain benefits directly, and then aggregate these values. But this would require collection of individual preference data.

Aggregation of Attribute Values

Unlike market goods, major regulations tend to offer large, indivisible bundles of effects that act on multiple attributes simultaneously and that cannot be separated or scaled down. Addition is appropriate for aggregating or reducing multiattribute effects to a single number only if the following conditions hold:

1. The value of a change in the level of one attribute is preferentially independent of the changes in any other attribute levels (Keeney and Raiffa 1976: Chapter 3).
2. The magnitudes of the changes in attribute levels brought about by the regulation are small enough so that decreasing (or increasing) marginal utility effects can be ignored.

The effect of these assumptions is that multiplying by k the change in each attribute's level caused by a proposed regulation multiplies by k the benefits attributed to the regulation. This is realistic only for marginal (i.e., sufficiently small) changes and only if the effects of bundling together changes in the levels of several different attributes can be ignored.

Example 3: Multiattribute health benefits from environmental regulation

The value of changing several attributes at once is assumed to be representable as the sum of the values of changing each one separately, holding the levels of the others fixed. This assumption would presumably not hold for a proposed emissions control regulation that would both (a) extend average life expectancy (e.g., by decreasing mortality rate from respiratory failure) and (b) make each year of life more pleasant by improving average ambient air quality and by reducing the frequency of respiratory illnesses. This assumes that life expectancy, average ambient air quality, and frequency of respiratory illnesses are among the relevant attributes used to describe the effects of the proposed regulation. When changes in the levels of different attributes complement each other as in this example, it is not realistic to combine the values of the changes in the levels of the different attributes through addition. Indeed, it is no longer possible to assess the benefits from changes in each attribute in isolation: the context of changes in other attributes must be considered simultaneously. Thus, a more complicated aggregation formula is needed to reduce the changes in the levels of multiple attributes to a measure of the benefit that an individual derives from these changes.

Aggregation across Individuals

Any approach to benefits assessment must face the fact that different individuals will receive different benefits from any proposed regulation. What the regulation produces is a *distribution* of benefits (and disbenefits) across individuals. The problem of aggregation is to reduce such benefit distributions to single equivalent aggregate benefits.

In the econometric approach, aggregation of individual preferences is usually done indirectly. Rather than actually eliciting individual preferences and then worrying about how to combine them, the surrogate prices are used to get from changes in attribute levels to changes in aggregate benefits. The problem of aggregation is solved implicitly in the process of choosing or identifying the surrogate price weights to be used.

The market solution approach

Ideally, these prices are assumed to result from (or to act as if they resulted from) an underlying market or market-like adjustment process that implicitly takes care of the aggregation of individual preferences. We will call this process (whose nature is often left rather vague) the *implicit market mechanism*. Just as equilibrium prices for market goods in simple partial-equilibrium models are determined by the intersection of aggregate (market) supply and demand curves, so surrogate prices are supposed to reflect aggregate equilibrium values for nonmarket goods and for the hypothesized implicit market adjustment mechanism, which does for these goods what the perfectly competitive market does for market goods. In effect, the problem of aggregation is assumed to have been solved by the market or its analog in producing the weights used to evaluate changes in attribute levels.

While this implicit market solution to the aggregation problem is often cited as providing at least a theoretical justification for the use of single weights rather than entire population preference distributions in assessing benefits, it suffers from three major flaws:

1. On the theoretical side, the notion of an implicit market for nonmarket goods is problematic. Making it precise for public goods leads to the so-called *Lindahl equilibrium prices* for public goods. A distinguishing feature of these prices, however, is that, unlike market prices, they are different for different individuals, depending on individual preferences. This point will be examined further in the discussion in Chapter 6 of the hedonic price method, which is one implementation of the econometric approach.

Similarly, the theoretical basis for pricing of public goods is well known: they should be produced at levels that equate the technologically determined marginal rates of transformation for production to the sum of individual marginal rates of substitution, as determined by individual preferences (see Chapter 2). Individual marginal rates of substitution for public goods may differ, since opportunities for the equalizing exchanges that are characteristic of market goods are not available. Thus, careful pursuit of the implicit market idea throws us back to the position of having to collect information about individual preferences.

2. On the practical side, aggregate market demand (or supply) curves even for true market goods may be very difficult to identify from

empirical data. In particular these curves are econometrically "under-identified" by any single market equilibrium data set on prices, quantities, and even local price elasticities.

Underidentification is most severe if the proposed regulation will cause large (nonmarginal) changes in the levels or prices of goods, so that observed aggregate demand relations must be extrapolated to unfamiliar situations. These difficulties also make estimation of aggregate benefits for nonmarket goods difficult if the surrogate prices for these goods are estimated (via the hedonic price method, or from cross-price relations with market substitutes or complements) from estimated market price data.

3. Finally, the use of aggregate demand curves to provide estimates of aggregate benefits presupposes that these curves adequately represent aggregate preferences. If the initial distribution of income is unfair, or if externalities are present, then the market demand curves may not be acceptable as guides for social decision making. In addition, individuals must have roughly similar preferences for tradeoff rates among market goods for the aggregate demand curve to be interpretable as representing aggregate preferences (Hildenbrand 1983; Bergstrom and Cornes 1983).

The policy judgment/sensitivity analysis approach

A common alternative to the implicit market approach is to leave the weights unspecified, as policy variables to be selected by the regulatory agency. More exactly, the range of values (weights) implied by different decisions is back-calculated and presented through statements such as "if the value of hearing impairment is less than $200,000 per case, then this anti-noise regulation should not be implemented: its benefits are outweighed by its costs. On the other hand, if the value is over $500,000, then the regulation should be implemented. For values in between, more information must be collected to allow an unambiguous determination to be made."

Presentation of results in this *sensitivity analysis* format is always desirable. It provides a way of dealing with some of the uncertainties that are inevitable in cost-benefit analysis. In the above example the decision maker is spared the difficult task of assigning an exact dollar value to hearing impairment; he must only decide whether it falls above, below, or within the interval from $200,000 to $500,000 per case. This sort of decision is in general much easier than choosing a dollar equivalent for the benefit of preventing a hearing impairment.

The sensitivity analysis approach is very useful for bypassing or reducing the problem of choosing surrogate price weights that adequately capture the aggregate benefits per unit change in attribute levels. It is particularly useful if there are no more than three attributes to be evaluated. With more attributes the relation between assigned values and resulting decisions may become too complex to provide much guidance (see von Winterfeldt and Edwards, in press, Chapter 11).

Even after sensitivity-based simplification, choice of an option still requires a policy judgment. Implicit in this whole approach to selection of surrogate price weights is the notion that ultimately some policy maker will make the required choice. In other words, there is a presumption that someone (supposedly representing the public interest as he sees it) will act as an intuitive synthesizer of public preferences. The aggregation problem is then solved holistically by this decision maker. Minor variations on this theme are to leave the choice of weights (or of decisions, implying a partial specification of the relative values of the weights) up to agency policy, or to infer them from past policy decisions. In either case, choice of weights is still seen as a question of policy choice by public decision makers rather than a matter of directly eliciting and aggregating individual preferences.

The drawback in this approach is that no objective basis in market or survey data is provided for public decision making. Moreover, the decision intervals or regions (i.e., the sets of surrogate prices corresponding to different decisions) are found through simultaneous consideration of the (partially evaluated) costs as well as the benefits of proposed alternatives.

Choosing the Unit of Benefits

Dollar values: willingness to pay and willingness to accept

The surrogate price weights for attributes, like ordinary market prices, are usually expressed in terms of the dollar value per unit change in attribute level. For many health and safety benefits, the attribute levels in question are actually probabilities of various undesirable consequences, so that dollar evaluations actually involve estimates of the dollar equivalents of small risk reductions. Implicit in such evaluations is a presumption that *dollar expressions of benefits are meaningful* in that they can be interpreted unambiguously and consistently by different users of a benefits assessment. This assumption has been much debated.

The price (or dollar value per unit) of a market good in a perfectly competitive economy has an exact meaning. It is the minimum amount of money that any supplier of the good will accept from any buyer in exchange for one unit of the good, since he knows that he can always find other buyers at that price. Simultaneously, it is the maximum amount that any buyer of the good will pay to any seller for a unit of the good, since he knows that he can always find other suppliers who will sell at that price.

The situation for nonmarket goods, such as the public goods provided by environmental regulations, is very different. The goods supplied by regulations can in general be obtained only through coordinated action, so that there are no competing suppliers. Individuals with different tastes cannot

5. Theory of Regulatory Benefits Assessment

costlessly adjust their personal consumption levels of public goods, such as cleaner air or tighter industrial safety standards for hazardous facilities, to reflect their personal tradeoff preferences. The nature of collective choice and of regulation is that tradeoffs are made, in deciding how to use collective resources, that everyone must live with. This means that objective prices, or exchange ratios between goods, arising from the decentralized equating of individual buyer marginal rates of substitution (determined by individual preferences) to individual seller marginal rates of technical substitution (determined by production costs) do not exist. Tradeoff ratios are set through policy decisions, not arrived at through decentralized production and consumption decisions.

These elementary facts imply that the surrogate prices used by the econometric approach to assess regulatory benefits in dollar terms do not have the objective equilibrium interpretation of true market prices in terms of voluntary exchange (or production and consumption tradeoff) ratios. Without such an equilibrium interpretation, however, it is not exactly clear what a dollar value for a benefit does mean.

The difficulties in interpreting the dollar values of nonmarket, as opposed to market, goods may be summarized as follows:

1. UNIQUENESS. For a market good (defined as a transferable commodity with the property that consumption/ownership by one agent costlessly precludes consumption/ownership by others), the maximum amount that an agent will pay for a unit of the good is the same for all agents and is in this sense objective. Even agents who do not enjoy consumption of the good would buy it at below-market price, if given the opportunity, for the purposes of arbitrage. For a nonmarket good such as national defense or clean air, however, different individuals would be willing to pay different amounts per unit obtained. Thus, buying prices are not the same for all individuals but follow some distribution across individuals. The buying price for a nonmarket good, in terms of the maximum amount that any individual would be willing to pay for it, is not unique.

2. DIVERGENCE OF BUYING AND SELLING PRICES. For market goods in a perfectly competitive economy, the selling price determined by technical production costs and the buying price determined by individual preferences and budget constraints adjust to each other until an equilibrium production and consumption level is reached, at which the two prices are exactly equal. For nonmarket goods no such adjustment takes place, and buying and selling prices may differ.

Since the benefits of health, safety, and environmental regulations are generally not owned or transferable, the selling price for such a good must be carefully defined. Usually it is defined as the least amount of money that someone would have to be paid to make him willing to forego his share in the benefit if he could do so, assuming that he was initially entitled to that share (so that in effect he is given it, rather than having to make or buy it). Thus, selling prices for nonmarket goods produced via collective action are defined, like buying prices, purely in terms of individual preferences. Like buying prices they can differ for different individuals.

Moreover, in contrast to market goods' prices under arbitrage, an individual's buying and selling prices for the same nonmarket good will in general be different. Indeed, under the assumptions of subjective expected utility theory, they can only be the same if either

> He is completely risk-neutral, so that he will always be indifferent between an uncertain prospect and its expected value; or

> His buying (and selling) price for a risky prospect does not depend on his wealth, so that he will be as willing to gamble when he has very little money as when he has a lot (Raiffa 1968).

Neither qualitative characteristic seems very plausible. In general, the selling price for a nonmarket good tends to substantially exceed its buying price even aside from possible income effects (Knetsch and Sinden 1984) so that the price of the good, even for a single individual, is not unambigously defined.

In empirical work, buying prices, or willingness-to-pay (WTP) values, are usually used in preference to selling prices. Buying prices are most appropriate for evaluating benefits (i.e., outcomes preferred to the status quo). Compensation prices (the least amounts of money that one would be willing to accept (WTA) to put up with undesirable changes if one had the choice) are generally used to assess disbenefits. The intuitive notion is that if the sum of the WTP values from those willing to pay for a regulation exceeds the sum of the WTA values for those who would pay to avoid it (plus any other costs of implementation), then under the potential Pareto improvement (PPI) rationale the regulation should be implemented. On the other hand, there is no logical reason why benefits should not be evaluated in terms of WTA compensation prices for those who benefit (how much would they have to be paid to compensate them for not having the regulation implemented?) and WTP values from those who do not benefit from the regulation (how much will they pay to avoid it?). The regulation would then not be implemented unless the former exceeded the latter.

Unfortunately, the decision on whether to implement a regulation under the PPI criterion can depend on which of these two approaches is used to represent individual preferences. As discussed in Chapter 4, the choice between them can be thought of as a policy decision about the initial allocation of rights. Are those who would gain from a regulation entitled to it, or are those who would lose entitled to not have it imposed on them?

5. Theory of Regulatory Benefits Assessment

Using WTP to measure benefits and WTA to measure losses reflects a policy judgment that the status quo assignment of rights is correct, so that those who would change it ought to be able to compensate those who would not.

3. THE COLLECTIVE NATURE OF REGULATORY BENEFITS. In a perfectly competitive market, each individual's production and consumption or sale and purchase behavior affects only his own welfare. He can thus assess his buying and selling prices for goods by considering only his own preferences and the prices and production costs available to him. In assessing the buying and selling prices for regulatory benefits, however, an individual must bear in mind that he is dealing with a collective good. In expressing a willingness to pay for national defense, for example, he cannot reasonably be asked to assess the value of his share alone; the good being evaluated is inherently collective.

Instead, he must consider how much of society's collective resources he would like to see allocated to this end and how much he himself would be willing to pay, given the amounts that others will pay. Indeed, how much he is willing to pay may depend on his beliefs about how much others are willing to pay – and, in particular, on how likely he thinks it is that the proposal will be implemented in the absence of his contribution. Not only does this raise the prospect of strategic misrepresentation of preferences by potential free riders (Green and Laffont 1979); but it also shows that expression of individual preferences for regulatory benefits in WTP terms is complicated by two factors:

> The willingness-to-pay question for individuals cannot be separated from the question of how the collective choice is to be financed (who will pay how much, and what collective tradeoffs are being made?).

> An individual's WTP cannot be completely characterized by a single number, as for market goods, but rather by a function relating his own WTP to the payment levels of others.

In summary, the meaning of an individual's dollar WTP value for a regulation is unclear unless we know his assumptions about the financing method to be used and the WTP's for other members of the population. Commonly proposed financing alternatives include

1. Having each person pay in proportion to his stated willingness-to-pay, with the proportion chosen so that total costs are just covered.
2. Having each person pay a fixed share of the cost, with the share size (but not the decision whether to implement the regulation) independent of his announced WTP.
3. Having each person pay his stated WTP, with any excess or deficiency being absorbed by the government.

These and similar alternatives have been discussed from a theoretical perspective on the basis of whether they provide incentives to accurately represent or to misrepresent preferences (Freeman 1979). In practice, of course, financing is not targeted for one specific proposal at a time, and so the question of what individuals assume about the social decision-making process in thinking about the maximum amount of money that they would want society to spend in implementing a proposed regulation, and in thinking about their own personal WTP values, remains open.

In addition to these conceptual ambiguities, there are a number of practical psychological difficulties with the use of monetary measures of individual preferences, especially for collective goods. Some of these have been reviewed in Chapter 4. For example, the risky option that an individual considers worth the most money is not necessarily the one that he will choose if a choice is given to him. Thus, from a psychological standpoint, the meaning of a dollar evaluation as an expression of individual preference, and hence the way that such evaluations should be used in guiding policy, remains somewhat vague.

Nondollar values: utilities

The social utility approach described later avoids some of the difficulties with dollar valuations of nonmarket goods by substituting a nonmonetary index – *utility* – as a numerical representation of individual preferences. Some of the methods described in Chapter 6 also use utility models, employing expressed preference data collected in interviews or questionnaires, as the basic representation of preferences to predict monetary WTP or WTA values. The basic strategy of the social utility approach to benefits assessment is consistent with the steps outlined at the beginning of the chapter. It calls for guided elicitation and representation of individual preferences and aggregation of changes in individual welfare (as defined by the preferences of those involved) into an overall measure of social benefit. The aggregation is accomplished through an explicit formula (usually weighted summation) reflecting ethical judgments, rather than through the market mechanism. This makes it attractive in settings where no well-defined market exists, or where it is felt that, because of maldistribution of income or imperfections in market structure, market prices fail to adequately represent individual preferences. The normative principles leading to weighted summation as a device for aggregating individual utilities to obtain social benefits are discussed later in the chapter.

Time, Risk, and Uncertainty

We have so far treated the problem of benefits assessment as an essentially static one in which announced individual preferences are used to guide the selection of one out of a fixed set of alternative policy options. Here we turn to the more realistic case in which benefits unfold and consequences of a particular policy choice become better understood over time. Instead of choosing an alternative and then living with the consequences ever after, the policy maker chooses a (possibly temporary) action, observes the consequences for a while, and then chooses another action. A *policy*, in this context, refers to an entire sequence of choices (or a rule for making choices) that is *conditioned* on information as it becomes available. Selection of an initial action need not commit one to a specific policy, and difficult decisions can sometimes be deferred until more information is available, although typically at the price of some waiting cost or penalty (cf. pp. 142–151).

Basic Strategies

Our basic approach to benefits assessment so far has always centered on aggregation of individual preferences, represented by individual benefit (e.g., WTP or utility) functions, to obtain or define social preferences, represented by a social welfare or aggregate benefit function. The policy alternative to be selected is then determined as the one that is expected to maximize (the increase in) net social benefits.

This prescription, however, is ambiguous if the consequences of implementing various alternatives (i.e., the resulting time streams of changes in attribute levels) are uncertain, and if different people have different expectations about the probable consequences of each alternative. The alternative that is expected by one person to maximize net social benefits may not be the one that is expected by someone else to do so, even if both parties agree on the definition of the social benefit function, for one person may be better informed than the other. Similarly, the policy that is expected by even one individual to maximize net benefits may change as more information becomes available to him. The result is that concepts such as Pareto efficiency and expected benefits must be defined with respect to specific information states (who believes what?) in order to be unambiguous (Holmstrom and Myerson 1983). Welfare analysis can only take place within a framework of information and beliefs (or assumptions), and this framework must be specified so that the results of the analysis can be unambiguously interpreted or adjusted by users with access to different information.

Three basic strategies for dealing with uncertainty about the consequences of policy choices can be identified:

1. In the *decentralized* approach each individual is asked to use his own beliefs and information and his own personal discount rate or time preferences to assess his subjective expected benefit (SEB) for each alternative. Of course, the value of any current collective choice may depend on what subsequent choices are made; the respondent may then be asked to use his own beliefs, however vague, about these subsequent decisions in deriving his SEB for each current alternative. Or each respondent may be asked to assess his SEB for alternative policies rather than for the current alternatives

alone. The resulting subjective expected benefits are then aggregated to obtain an aggregate expected benefit for each alternative.

2. The *market* or institutional approach to dealing with uncertainty is similar to the decentralized approach, except that instead of submitting subjective expected benefits (utilities or WTP values) to an analyst for aggregation, individuals are expected to take advantage of market mechanisms such as insurance pools, contingent contracts, commodity options, and so forth to deal with risk and uncertainty through decentralized actions. In other words, market mechanisms or similar deliberately designed institutions or processes may to some extent substitute for regulatory control as devices for managing risk and uncertainty. Where such mechanisms exist and operate efficiently, they may both provide market data and reduce the need for benefits assessments in support of regulatory control. However, these process approaches fall outside the scope of this discussion and in any case fail to address many of the health, safety, and environmental risks that are currently managed through regulation. We will therefore not consider them in detail.

3. In the *centralized* approach, respondents are asked to assess the values of different time streams of attribute values, or are asked to assess utilities for specified probability distributions over these time streams. In other words, they are asked to provide preference data for consequences or probable consequences but not to make judgments about what the probable consequences of (i.e., time streams of resulting changes in attribute values from) different policy decisions might be. This is instead done centrally, using analytic models and the expertise of specialists. Thus, preferences for consequences are separated from technical expertise. Centralized information on consequences is combined with the aggregate social benefit function for uncertain consequence streams to obtain a centralized expression for the expected net social benefit of each alternative.

Comparison of Basic Strategies

The market approach is most compatible with holistic preference elicitation techniques, in which actions rather than consequences are evaluated directly by respondents. The decentralized approach is more compatible with the multiattribute approach and has the important advantage that it does not force a choice among alternatives which respondents may lack the information to evaluate confidently. Holistic evaluation methods in effect force respondents to make a choice among alternatives (or to evaluate alternatives, depending on the type of implementation) at the time of the survey. However, one of the advantages of living in a dynamic world is that such decisions can often be deferred while additional information is collected that will ease the choice and increase the expected utility of the final decision (by diminishing the probability of error.) Thus there can be a real cost in terms of reduced individual expected benefits from basing policy choices on forced holistic evaluations made without full information.

In summary, we believe that the centralized strategy for dealing with uncertainty is preferable, in that it allows specialization and expertise to be brought to bear on empirical questions, while still using individual preferences to determine values and choices. We will therefore concentrate on this approach. The main remaining problems have to do with representing and aggregating individual preferences for uncertain time streams of consequences (changes in attribute levels); approaches for solving these problems are discussed below. The question of what time stream of consequences will in fact be produced by a particular policy can be separated out of the analysis, as can the social decision problem of choosing a policy.

Sequential Decision Making and the Process Perspective

Prospective regulatory benefits assessment is primarily useful in a *choice from a set* decision context, where it is necessary or desirable to decide which of a few competing alternative regulatory approaches to implement and where the choice determines a subsequent flow of consequences. The basic logic of the approach is to estimate the aggregate net WTP or probable consequences of each alternative, assess the benefits (e.g., the social utility) corresponding to each consequence set, and then to choose the alternative giving the greatest net benefit.

When the matter of choice is not a single decision from a limited set of alternatives but rather the first in a sequence of wholly or partially reversible decisions with long-deferred and gradually revealed benefits, then the choice-from-a-set paradigm is no longer appropriate, and benefits assessment techniques are apt to be inefficient as guides to action. The reason, as discussed above, is that decisions in such cases must be coordinated over time, and the value of a current decision can only be assessed in the context of the overall policy of which it is a part. Time streams of consequences are usually generated by overall policies or processes, instead of being irrevocably determined by the current regulatory decision. And a regulatory policy will in turn often be part of an overall regulatory package, so that the benefit from a particular regulation can only be assessed in the context of the policy package to which it belongs. The proper question for benefits assessment is therefore not whether it is possible to find a Pareto-superior (or potentially Pareto-superior) alternative to a particular proposed regulation but whether it is possible to find a Pareto-superior package of policies that could be implemented instead.

From this perspective, benefits assessment based on consequence analysis of one proposed regulation at a time must be seen as providing only limited guidance for policy makers. Indeed, when health, safety, or environmental consequences flow from a coordinated set of regulatory policies administered by different agencies, it is difficult to define or isolate the share of consequences that is properly attributable to any single policy or regulation – especially if there are substitute or complement effects among them.

These considerations suggest that the usual regulatory benefits-assessment paradigm outlined at the beginning of this chapter, although comparatively well suited for one-shot, essentially static (i.e., choice-from-a-set) decisions, is not sufficiently rich to adequately guide dynamic decision processes and policies in which the timing of implementation activities and the decisions of future decision makers are important. An alternative framework is needed to analyze such cases. One possible alternative that has already been mentioned is the *process approach* (see Chapter 3). This approach, as described, for example, by Kunreuther, Linnerooth, and Vaupel (1984), makes the regulatory process, rather than individual regulations, the object of analysis. Just as the perfectly competitive market mechanism can be analyzed from the standpoints of efficiency, equity, and political acceptability, without ever asking about the absolute magnitude of the benefits that it generates, so deliberately designed institutions, including regulatory ones, can be analyzed and improved without using the information provided by conventional benefits assessments. Key questions would center on whether opportunities for informed individual participation are available, whether individual preferences for collective actions are expressed and acted on, whether opportunities for mutually beneficial transactions are discovered and efficiently exploited, and, especially, whether the process could be improved to yield a Pareto–superior overall performance. The emphasis in this approach is on designing processes that maximize benefits rather than on identifying explicit actions whose selection will maximize benefits.

In summary, benefits assessment of individual regulations encounters several methodological difficulties when risks and uncertainties about consequences are involved. Risky prospects or regulations with uncertain consequences are often best assessed in the context of the portfolio to which they belong. Indeed, analyzing the benefits from such prospects in isolation, without considering the pattern of correlations across their outcomes and across outcomes to different individuals, is apt to be misleading. Attempting to assess regulatory benefits within the context of related, possibly correlated decisions, however, leads to consideration of the overall decision process as the unit of analysis.

A Comparative Evaluation of Alternative Approaches

To help put the many approaches and techniques of benefits assessment for nonmarket goods dicussed in this study into perspective, Figure 5–1 presents a somewhat simplified overview of the whole set of alternatives. At each branch point in this tree there is a choice of approaches. The tree ends in a set of specific methods that are discussed individually in Chapter 6.

5. Theory of Regulatory Benefits Assessment

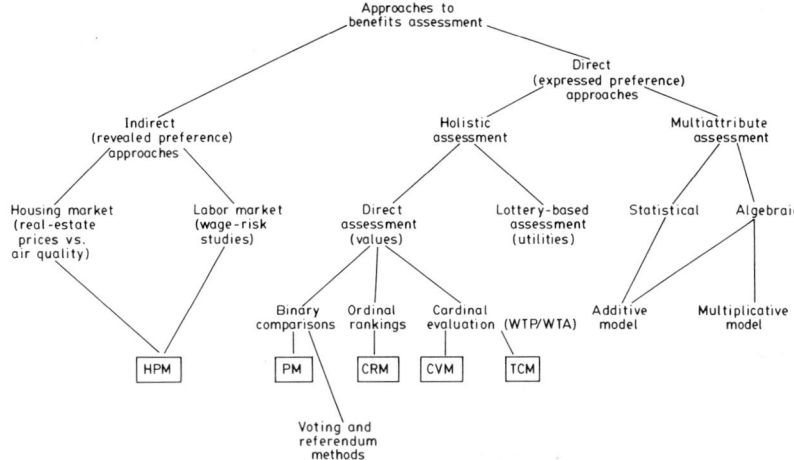

Fig. 5–1. Approaches to benefits assessment for nonmarket goods (see also Chapter 2).

These methods can be thought of as resulting from different sequences of choices about what approach to take to benefits assessment.

In this section we consider how the choices at the different branch points in Figure 5–1 might be made and examine the implications of each choice in terms of data requirements and study design. The motivations, theoretical bases, and limitations of the approaches in Figure 5–1 are reviewed above and in the latter half of this chapter. The specific methods at the bottom of the tree are treated in Chapter 6. We concentrate here on general comparisons, designed to provide overall guidance on the choice of a path through the options presented. These comparisons are organized as a series of comments for making choices at each level in Figure 5–1.

Revealed Preference Approaches versus Expressed Preference Approaches

Both revealed-preference, market-based approaches and expressed-preference, questionnaire- or interview-based approaches to regulatory benefits assessment have the same fundamental goal: to identify which among a set of proposed regulatory alternatives will lead to the greatest increase in social welfare. Both agree that social welfare should be defined in terms of a (probably additive) aggregation of individual welfares and that individual welfares should be defined in terms of individual preferences. Aside from these fundamental points of agreement, however, different methods can differ even in principle about how individual preferences should be measured and about the economic nature of nonmarket goods, and hence many differ in the types of data they require and in implementation strategy. Chapter 6 contains several specific examples.

The econometric (revealed-preference) approach assumes that actual or implicit market prices are the proper measure of individual preferences. The basic data requirements therefore focus on obtaining market prices for suitable markets (e.g., housing or labor markets). Actual market prices are used to deduce implicit prices for nonmarket goods, such as air quality, using regression analysis or related econometric techniques. If necessary, in the absence of adequate market data, surveys may be used to obtain announced WTP values from which implicit prices can be calculated. The value of a regulatory benefits package is then found by pricing the changes in attribute levels that the regulation would cause and summing the resulting dollar values.

Expressed-preference and psychometric approaches proceed from the assumption that the proper measure of individual preferences is choice behavior among alternatives. When the alternatives involve public or other nonmarket goods, prices do not convey the required information about individual preference tradeoffs, and surveys or interviews must be used to gather preference data (usually involving hypothetical or potential choices) directly. In its most radical form, this perspective not only asserts that missing price data must be substituted for with collected preference data but goes on to say that monetary measures of individual welfare are inappropriate, because

1. They needlessly confound personal preference data with personal budget constraint and perceived asset position data;
2. They rely for clear interpretation on a model of optimization (in which prices emerge from the first-order conditions for constrained optimization over a continuous set of alternatives) that is inappropriate for choices involving discrete, indivisible public goods and other sorts of nonmarket goods; and
3. The equilibrating exchange mechanism that allows a single set of prices to represent everyone's preference tradeoff rates for market goods in equilibrium (after everyone has adjusted his own personal consumption bundle accordingly) does not exist for indivisible public and other nonmarket goods, so that single aggregate implicit market prices lack any natural equilibrium interpretation and are not suitable as a basis for measuring welfare.

The econometric approach requires market data and avoids the need for surveys when such data are available. It is based on the use of revealed rather than announced preferences to the extent possible. Expressed preference approaches require surveys and therefore use announced rather than revealed preferences. Both sets of approaches use regression techniques and related tools to smooth and idealize raw preference data.

To overcome the theoretical criticism that it uses an inappropriate optimization model (since prices typically do not arise in optimization over discrete choice sets), the econometric approach must assume that the opportunity set

5. Theory of Regulatory Benefits Assessment

that individuals choose from contains an extremely rich and diverse set of alternatives (which can be modelled analytically as at least approximately convex). Richness in the choice set and diversity in individual choices are also necessary to overcome the econometric problems of estimating implicit prices. To the objection that aggregate prices are not meaningful measures of welfare for public goods, there is no adequate response: it is generally admitted that data at the level of individual preferences (e.g., individual marginal WTP values at different levels of a public good) are needed to support efficient decision making about the levels of public goods (see, e.g. Bergstrom and Cornes 1983).

To overcome such theoretical difficulties, the econometric approach can interpret monetary values such as announced individual WTPs, not in terms of prices or exchange ratios, but as direct expressions of value or preference intensities. That is, dollars can be used as the units on an individual's value scale. In this case, however, using a value or utility scale calibrated directly in terms of preferences for alternatives rather than abstract dollar values seems more natural, and is in fact required for consistency with the axioms of rationality, as is discussed in the second part of this chapter.

Holistic versus Multiattribute Assessments of Preferences

An important and difficult task in both the econometric and survey-based approaches is to figure out what attributes people take into account (or would ideally like to take into account) in choosing among alternatives. Although many different sets of attributes can typically be used to represent a given set of relevant constructs, failing to represent one or more important and relevant constructs in the description of alternatives can lead to erroneous policy conclusions. Specific examples of the sorts of attributes that should probably be included in the description of alternatives, but that are often omitted (since they would not be needed or would not even be relevant for traditional decentralized market decison making) include:

• Specification of the financing scheme or payment vehicle to be used to obtain the alternative (who will pay what if the alternative is adopted?). Examples could include sales taxes and surcharges on certain activities.

• The distribution of changes in attribute levels across different stakeholder groups if the alternative is adopted (who else will be affected and how?).

• The preferences of others regarding alternative attribute levels.

In addition, the distinction between private decisions (what is the greatest amount that you personally would be willing to pay for this benefit?) and collective ones (what is the greatest amount that you think society should pay for this benefit?) must be preserved during the data collection.

Once the set of attributes to be used in describing alternatives has been chosen, a decision must be made about whether alternatives as wholes, represented as bundles of changes in attribute levels, should be evaluated (the holistic approach), or whether changes in individual attribute levels should be evaluated and then aggregated (the multiattribute approach). Although not shown in Figure 5–1, this choice must be made in the econometric as well as in expressed-preference approaches. The holistic approach may be preferable if only one or a few attributes (and underlying constructs) are involved. However, for regulations affecting many attributes – all of them important – announced respondent preferences are apt to be confused and variable, and a multiattribute approach would then be preferable.

The holistic approach has the advantages of simplicity in design and concreteness in the questions asked. Each respondent is asked about his preferences for specific, real alternatives. On the other hand, if preferences among the alternatives are quite uncertain because of the complexity of the stimuli and the difficulty of the choice task, a multivariate statistical analysis may be required to discover the underlying true preference model, which will itself be expressed in multiattribute form. The multiattribute approach deals with preferences over attribute levels directly, rather than inferring them from statistical analysis of noisy holistic preference response data, and may therefore require a much smaller sample to determine an adequate model of preference. On the other hand, it requires careful design of the survey or questionnaire and asks more abstract and hypothetical tradeoff questions, and is thus further removed from the reality of direct preference data for actual alternatives.

The choice between holistic and multiattribute preference evaluation methods must ultimately be made on the basis of the homogeneity of preferences in the population, the difficulty and unreliability that respondents experience with the holistic questions, and the cost per interview of each method. The holistic method saves on interview and stimulus design costs but may cost more at the analysis stage.

Level of Measurement

The tradeoff between complexity in interview design and implementation and complexity and cost of analysis continues in deciding whether to ask for binary preference comparisons, whole preference orderings, or cardinal, numerical assessments. In general, choosing a simpler response mode (e.g., binary responses) improves the reliability of each respondent's answer and reduces the cost per interview. On the other hand, statistical analysis of the response data to obtain an underlying cardinal multiattribute model is correspondingly more difficult and requires a larger sample size to fit the model.

The specific data requirements, assumptions, advantages and disadvantages, and domains of application for each of the individual methods at the bottom of Figure 5–1 are considered further in Chapter 6.

The Social Utility Framework for Benefits Assessment

The econometric approach outlined earlier in the chapter represents what is probably the dominant present paradigm for benefits assessments in regulatory agencies. The limitations noted for this approach are, to a large extent, limitations in the state of the art.

In the past ten years or so, however, an alternative approach has been gaining increasing importance. We will call it the *social utility* approach to benefits assessment. Based on the theory of utility functions for representing (idealized) individual preferences under uncertainty, and on the theory of social welfare functions for aggregating individual preferences into measures of social benefits, the social utility approach has in recent years been evolving from an esoteric academic tool of theoretical decision analysis to a widely applied tool for assessing and expressing the values of health, safety, and – to a lesser extent – environmental benefits.

Whether the social utility approach has matured sufficiently to serve as a viable alternative to the econometric approach is a matter on which professional analysts disagree. Indeed, some practitioners question the usefulness and soundness of the method under any conceivable practical circumstances, while others insist that it has already proven itself repeatedly in a variety of practical applications. Insofar as it provides new perspectives on the proper treatment of risks, on methods for expressing and aggregating individual preferences, and on distributive equity under uncertainty, however, the social utility approach cannot be ignored in any up-to-date discussion of methods in benefits assessment. Moreover, the recent proliferation of applications to case studies involving life, health, and safety effects or risks brings it squarely within the domain of potentially useful approaches for satisfying the concerns of Executive Order 12291. For this reason we spend the rest of this chapter describing the principles and limitations of the social utility approach.

Representing the Preferences of Individuals: Market Goods and Nonmarket Goods

In principle, the market representation of individual preferences through prices reflects individual choice behavior under resource constraints. Given that an individual cannot afford every consumption bundle that he might want, the market forces him to reveal his true preferences by making him choose among the alternative bundles available (i.e., those that satisfy his budget constraint). His choice thus reflects both his tastes and his income or wealth constraint.

Representation of individual preferences in the social utility model is based on exactly the same principle of constrained choice behavior. But the set of feasible consumption opportunities from which the individual chooses can be more general than the set of market bundles that are consistent with his budget constraint. For example, each individual might be asked to express his preferences for alternative levels of a public good, corresponding to choices of a collective consumption bundle (and resource commitment). Therefore, the utility approach does not try to represent individual preferences through prices, since choices may have to be made among indivisible, discrete regulatory alternatives or among other types of nonmarket goods, and since, in the absence of market adjustment mechanisms, income-constrained preferences may not be the most relevant guides to action.

Instead, utility theory takes the position that the true measure of individual preferences is the tradeoff that an individual is willing to make among alternatives – whether those are alternative affordable market bundles, alternative proposed regulations, or alternative public projects and financing schemes. The value of a good or of an alternative to an individual is defined by his willingness to trade it off against other goods or alternatives.

For continuously divisible, privately owned market goods, this trading off of goods against each other until their tradeoff ratios, or relative marginal utilities, have been equalized is accomplished through prices. Relative values can therefore be expressed in monetary terms. (Note that it is tradeoff ratios and not absolute values that are measured in monetary terms; money does not provide a value scale but only a medium through which exchange ratio tradeoff rates among goods can be expressed.) For indivisible public goods or other discrete, nonexchangeable alternatives, however, some other approach must be used to measure tradeoff ratios, for there is no market in which actual trades take place.

The approach taken by utility theory to the measurement of the relative values that an individual assigns to different alternatives is to look at the rates at which he will *risk* trading one option for another. The value of an alternative is defined in terms of how averse the individual is to a risk of losing it. Willingness to accept the risk of losing something (in exchange, of course, for the possibility of winning something better) provides a universal medium for expressing value, even in situations where there is *no market*, no possibility for subdividing or exchanging alternative goods, and hence no clear interpretation for prices or monetary measures of relative values.

More exactly, the utility proposal for measuring individual values is this. Suppose that an individual is asked to evaluate various regulatory (or other) alternatives. The first step is to have him rank these alternatives from most to least preferred, thus establishing an ordinal scale of preferences (see Chapter 4). It is assumed that he is able to do this. Next, the endpoints of this scale are assigned numerical values; for example, the least-preferred is arbitrarily assigned a value of 0, while the most-preferred is arbitrarily assigned a value of 100. These two numerical assignments are used to

5. Theory of Regulatory Benefits Assessment

calibrate the value scale; the values of the remaining alternatives are to be assessed by reference to these two endpoints.

Naturally, if new alternatives are introduced that fall outside these endpoints, then the value scale must be rescaled to include them. To avoid this inconvenience it is usual to choose as endpoints alternatives – possibly hypothetical – that are likely to bound all the actual alternatives.

Informally, the rest of the preference representation process consists of assigning to the remaining alternatives numerical values which both (a) preserve the preference ordering of alternatives (so that they all lie between 0 and 100), and (b) contain some information about the relative *intensities* of preferences for different alternatives. In the next subsection, we examine several specific theoretical rationales for making this assignment. The following section will then consider practical methods used to implement these theoretical approaches.

Theoretical Bases of Nonmonetary Value Measurement for Individuals

Direct assessment

One way to obtain numerical measures of preferences is to ask individuals to provide direct assessments of alternative values, using the values of the least-preferred and most-preferred alternatives (i.e., 0 and 100) as reference points. The respondent assigns to each alternative X a number, which we will denote by $V(X)$, reflecting that alternative's perceived distance from the two endpoints in terms of relative desirability. This is conceptually similar to having him directly assess abstract willingness to pay for different alternatives without specifying the tradeoffs implied by different answers (e.g., whether the good would be provided anyway). It can perhaps provide useful information, but lacks an underlying tradeoff-based rationale and interpretation.

Direct assessment, whether done on a monetary or on an abstract value scale of the sort we are discussing, requires an assumption that values actually exist on some sort of natural scale which can be translated by the respondent into numerical (dollar or value scale) terms.

Assessments using preference differences

An alternative, more formal, approach to value assessment is based on the notion of preference *differences*. Namely, given any four alternatives W, X, Y, and Z such that the respondent prefers W to X and Y to Z, it is assumed that he can say whether he prefers W over X *more strongly* than he prefers Y over Z. That is, it is assumed that he can rank preference differences, or strengths of preferences, as well as ranking alternatives. If this is possible, then under some mild technical assumptions, the value number $V(X)$ corresponding to each alternative X can be uniquely determined (once the value scale has been fixed) from the respondent's rankings of alternatives in terms of his preferences and of pairs of alternatives in terms of preference differences (Krantz et al. 1971). (By fixing the scale, we mean choosing numerical values to represent the values of the most- and least-preferred alternatives. Common choices are 100 and 0, but there is no logical reason why we cannot rescale the values to run from 0 to 10, or 0 to 1000, or $^-5$ to $^+5$.) The choices of origin and scale are arbitrary, but once they have been made, the rest of the values assigned to different alternatives should be uniquely determined.

The numbers $V(X)$, $V(Y)$, . . . represent the respondent's preferences, in the sense that for any four alternatives W, X, Y, and Z,

$V(W) > V(X)$ if and only if alternative W is preferred to alternative X;

and

$V(W) - V(X) > V(Y) - V(Z)$ if and only if W is preferred to X more strongly than Y is preferred to Z.

The representation $V(X)$, which embodies the assumption that strengths of preferences exist and can be meaningfully compared, is sometimes called a *measurable value function* (Dyer and Sarin 1979). Preference differences and measurable value functions are useful not just because they suggest preference elicitation methods based on ordinal comparison data (Horsky and Rao 1984), but also because they provide an intuitive interpretation of the numbers on a value scale. The use of measurable value functions in the actual elicitation and representation of individual preferences will be considered later, after the meaning of the abstract (i.e., nonmonetary) value scale has been been further explained.

Utility assessments using hypothetical lotteries

A third way of obtaining numerical representations of individual preferences provides a tradeoff-based interpretation for values on the value scale using the risk-tradeoff idea previously mentioned. To evaluate alternative X, the respondent is invited to consider a choice between (a) getting X for certain, and (b) playing a lottery that gives his *most*-preferred alternative with some probability p, and his *least*-preferred alternative (which in many applications, after a preliminary screening of the alternatives to be evaluated, will be the status quo) with probability $1 - p$. The reasoning is that if p is great enough (and certainly if $p = 1$), then the respondent will prefer the lottery to the alternative X; while if p is small enough (and certainly if $p = 0$), then he will prefer X to the lottery. As p is slowly increased from 0 to 1 therefore – so the argument goes – there must come a point, say p^*, where the individual will be exactly *indifferent* to either getting X with certainty or taking the lottery offering probability p^* of winning his most-preferred outcome (and probability $1 - p^*$ of winning nothing, i.e., of receiving the least-preferred outcome). This *indifference probability*, p^*, expressed as a percent, may then be defined as the *utility* of alternative X. It is denoted by $U(X)$.

The utility $U(X)$ of an alternative X provides a definition and measure of the value of X to the individual. If X is nearly as desirable as the most-preferred alternative, then $U(X)$ will be close to 100. If X is hardly better than the status quo (or least-preferred outcome), then $U(X)$ will be close to 0. However, it is not necessarily true that the utility value $U(X)$ must coincide with the measurable value $V(X)$ that would result from an elicitation based on preference differences. Lottery-based elicitations of the sort just described are sensitive to the *risk attitudes* of the respondent, while deterministic elicitation methods, such as those based on strengths of preference, presumably are not.

However, in many practical applications, it can be shown that $U(X)$ and $V(X)$ should in principle coincide. This fact, which is discussed further in the subsection on multiattribute utility assessment, is extremely useful and important for practical work. It means that the value scale can be interpreted in terms of both risk-based tradeoffs and of strengths of preferences. The two interpretations give identical answers for the value of an alternative under a wide range of conditions that have been found to hold, at least to a satisfactory approximation, in a variety of practical applications.

The utility concept: theory

The concept of utilities as measures of value is important less as a practical device for actually quantifying real preferences (although it is sometimes used for that purpose) than as a theoretical device for providing a clear-cut, tradeoff-based interpretation of the numbers on a value scale. Namely, $U(X)$, whether it is obtained by direct assessment, by hypothetical lotteries, or in some other way, may be interpeted conceptually as the rate (probability) at which an individual is willing to trade off getting X with certainty against a chance of getting his most-preferred alternative. This sort of tradeoff reflects preferences through constrained choice (where the constraint is that he cannot have both the alternative and the lottery), but is independent of the individual's monetary income or wealth, except insofar as these affect his preferences among lotteries and alternatives.

The fundamental assumption of utility theory is that individual preferences for alternatives, and for lotteries involving alternatives, exist and are sufficiently well structured so that numerical representation of them is possible. In addition, they are well enough structured so that the numerical representation is unique, once the value scale has been fixed. (As previously mentioned, by fixing the scale, we mean choosing numerical values to represent the values of the most- and least-preferred alternatives.)

The assumptions that individual preferences exist and that they are so tightly structured that they can be measured numerically, once an appropriate scale has been specified, are extreme idealizations. Empirical research (e.g., Hershey, Kunreuther, and Schoemaker 1982) shows that the utility values $U(X)$ elicited from an individual for various alternatives X depend very much on the elicitation procedure used. For example, direct assessments of the values of alternatives, with no consideration of lotteries, may produce different results from lottery-based evaluations – leading many authors to distinguish between *utility functions* (lottery-based) and *value functions* (deterministic). Similarly, lotteries in which the worst possible outcome is the status quo may produce different utilities from lotteries in which the worst possible outcome is better than the status quo. These limitations cast doubt on the reliability and usefulness of lottery-based assessments, and raise the question of whether utilities really exist and can serve as a useful basis for guiding public decisions.

The answer is that utilities do not really exist any more than coherent preferences do (see Chapter 4). Both are idealizations. However, both concepts are useful in guiding public decisions. The reason is that if individuals agree with and accept in principle certain axioms of rational decision making that they would *like* their preferences to obey, then it can be shown that they should act as if their preferences were represented by utilities (Luce and Raiffa 1957). In fact, they should act so as to *maximize subjective expected utility*, where utilities are defined by lottery-based tradeoffs, as described earlier, and subjective probabilities are defined similarly (in terms of indifference judgments between appropriately constructed lotteries).

The full development of *subjective expected utility* (SEU) theory is quite technical and involved (Roberts 1979). It provides the main intellectual framework for economic theories of rational individual choice under uncertainty. For the purpose of regulatory benefits assessment, there are two significant features of the theory, described below.

1. *Subjective expected utility is a normative, not descriptive, theory*

It has been convincingly demonstrated that real people do not act or make choices in ways that can be explained by SEU theory. Indeed, many individual choices are flatly inconsistent with maximization of SEU. However, insofar as the theory does prescribe how individuals would *like* themselves to behave if they wished to act in accord with certain abstract axioms of rationality, it provides an approach to defining idealized preferences that are appropriate for use in guiding public decision making.

According to this view, individual preferences elicited through questionnaires or revealed through actual choices may contain errors, inconsistencies, and biases. SEU provides a way of seeing through these confused responses to the true underlying preferences, much as a regression model provides a way of seeing through noisy responses to a simple underlying relationship. In both cases, it is assumed that a true underlying relationship exists; that is, that people may in effect be mistaken about their own preferences in answering particular questions. Added to this is the important assumption that people would, if such mistakes were pointed out to them, and after sufficient thought, admit that they had been mistaken and revise their preferences to bring them into consistency with the axioms of SEU theory for rational individual behavior. If these assumptions are accepted, then SEU provides a theoretical basis for fitting simple models to represent idealized individual preferences. These idealized preferences are then used as the basis for defining and measuring aggregate social benefit.

2. *Uncertainty is treated through subjective expected utility*

In addition to providing a normative model for individual preferences and decision making, SEU theory shows that if individual values for different alternatives are expressed in terms of utilities (rather than money or other physical units), then the value of an uncertain prospect is just its subjective *expected* utility. The practical implication is that individuals can be asked to evaluate uncertain prospects – for example, measures such as a mandatory seatbelt requirement or a ban on the use of a suspected carcinogen – whose true effects may be very uncertain. The resulting values, interpreted as subjective expected utilities, are then measured on the same scale as alternatives with known consequences. Thus, certain and uncertain prospects can be evaluated in a common framework, and their values can be assessed in the same terms. This is in sharp contrast to value scales that are calibrated in monetary or other physical units, where uncertain prospects have values that must be represented by unwieldy probability distributions or by certainty equivalents that are in general ambiguous (Krantz *et al.* 1971, Sec. 8.4).

Despite this theoretical advantage, we must reiterate that real preferences and subjective probabilities systematically violate the assumptions of SEU (Kahneman, Slovic, and Tversky 1982), so that considerable care must be taken in eliciting individual preferences, especially for uncertain prospects (i.e., alternatives with uncertain outcomes). The chief merit of SEU theory is therefore its contribution to our understanding of the meaning of a value (or utility) scale calibrated in terms of relative preferences.

The preceding discussion has reviewed the units used in the social utility approach to represent preferences: the value of each alternative is to be assessed relative to the values of two endpoint alternatives whose values – generally defined as 0 and 100 – implicitly calibrate the rest of the scale. Taking the concept of a value or utility scale for granted, we now turn to methods for assessing individual preferences on such scales.

Eliciting Preferences for Alternatives

Two approaches to value or utility assessment – direct assessment and hypothetical lotteries – have already been mentioned. Each has been considerably refined and tried out in different variants. For example, hypothetical lotteries may be used to discover certainty equivalents rather than indifference probabilities, by varying the prizes rather than the probability of winning (e.g., Hershey, Kunreuther, and Schoemaker 1982). Indeed, there is an entire battery of techniques available for estimating various sorts of utility and value functions (see e.g., Fishburn 1967; Currim and Sarin 1984). Considerable empirical evidence on the ability of respondents to use these methods coherently and reliably has also been collected, and useful rules of thumb – for example, that respondents can typically only identify indifference probabilities to within about ± 5 percent – have been developed (von Winterfeldt and Edwards, in press). Thus, there is a wealth of documented experience available to guide practitioners in the elicitation of individual preferences.

Eliciting coherent individual preferences can, in some cases, be difficult and time-consuming, requiring many iterations and interaction between the respondent and the eliciter. The analyst may have to provide considerable guidance to the respondent by pointing out inconsistencies and logical consequences of announced preferences. This is particularly likely to be the case where the respondent is unfamiliar with the alternatives that he is being asked to evaluate, and where each alternative has several attributes.

While such interaction and iteration are often useful in helping a single client to clarify his thinking in making an important decision – the context for which decision-analysis techniques were originally developed – they are generally not appropriate in surveys designed to collect preference data from a large sample of respondents. It is also not clear to what extent such interactive elicitation really elicits preferences and to what extent it creates them. Moreover, as a general rule of thumb, direct holistic assessment of values on a scale is often about as accurate and useful as – and much simpler than – more complicated, lottery-based methods (von Winterfeldt and Edwards, in press). For survey work, therefore, the key question is whether respondents can provide confident, coherent, direct assessments of the alternative(s) being evaluated on an appropriate value scale. (The problem of coordinating individual preference scales is considered below.)

In some applications, direct intuitive assessments may indeed prove satisfactory. Even then, design of a successful survey instrument requires considerable care. For example, a value scale running from 0 to 100 may produce different answers from one scaled to run from -5 to 5. There is a considerable literature on the design of survey instruments which will avoid or control for this and other response-mode biases.

In the domain of health, safety, and environmental regulations, however, direct assessment may not prove satisfactory as a device for eliciting preferences, no matter how skillfully the value scale is constructed and interpreted. The effects of most proposed regulations are too complex, diverse, uncertain, and insufficiently familiar to respondents to permit confident holistic value assessments. Even if the analyst bullies respondents into providing value assessments, it is questionable how much such assessments mean and whether they should be used as a basis for policy decisions. This restriction – that personal evaluations of unfamiliar and complex measures, about which the respondent may not feel he can make confident judgments, are not necessarily very useful or usable to policy makers – applies equally well to holistic judgments of willingness-to-pay as to holistic judgments expressed in terms of a value scale. Preferences expressed for poorly understood alternatives that the respondent has not thought carefully about and that he may not have been familiar with are not a satisfactory basis for public decision making, no matter how they are expressed.

The Multiattribute Approach

To get around the problem of asking people to express preferences among complicated, unfamiliar proposals, an indirect strategy may be used that is very similar in spirit to the econometric approach. In this approach the anticipated effects of each proposal are described and represented in terms of changes (or potential changes) in the levels of a number of descriptors or attributes. Surveys are then conducted to discover preferences for changes in attribute levels. This is similar to the econometric approach's attempt to discover or impute monetary values for changes in each attribute level, except that the values are now expressed in terms of value or utility functions for attribute levels rather than in terms of implicit prices or willingness to pay.

More exactly, the multiattribute strategy is as follows:

1. Each proposed alternative X, instead of being evaluated holistically, is first represented by a set $(X_1, X_2, \ldots, X_n)$ of changes in attribute levels that are expected to be caused by the proposed alternative if it is adopted. Here X_i denotes the change in the level of the ith attribute (e.g., expected number of accidents prevented or lives saved per year). If the effects of a proposed alternative are uncertain, then the analyst may work with a probability distribution over changes in attribute levels.
2. A survey is conducted to determine the values or utilities that respondents assign to changes in attribute levels. Specifically, a utility function $U(X_1, \ldots X_n)$ is assessed for each respondent. (We will assume that utilities rather than values are assessed; as previously indicated, the distinction can often be ignored, especially if there is no

uncertainty in the alternatives to be evaluated.) This function assigns to each set of changes in attribute levels a corresponding utility number on the value or utility scale.

There are two fundamental approaches to the assessment of utility functions:

The *statistical approach* presents respondents with a carefully designed set of stimuli (e.g., cue cards) for which they are asked to express preferences (e.g., by ranking or numerically rating them). These stimuli are constructed according to the usual principles governing the design of experiments (factorial designs; see, e.g., Montgomery 1976), with the attributes playing the roles of the factors in the design. The responses obtained – the preference data – are then analyzed through regression-based methods such as MANOVA, or similar statistical techniques, to produce a utility function which is in some sense most consistent with the responses obtained.

In the *algebraic approach* utility functions are assessed by eliciting preference (or, more exactly, indifference) judgments from the respondents. These correspond to equations constraining the utility function that is to represent the respondent preferences. The utility function itself is found by solving the resulting system of equations.

There is some evidence that the statistical approach is preferable to the algebraic approach in some marketing contexts. For a comparative assessment and fuller explanation of the two approaches and of the specific techniques for implementing each, the reader should consult Currim and Sarin (1984) and the references therein.

3. The changes in individual utilities corresponding to each proposed alternative being evaluated are then aggregated across individuals to obtain a measure of aggregate social benefit from the proposal. This aggregation phase is discussed later in the chapter.

Advantages of the Multiattribute Strategy

This strategy solves two problems that arose in the econometric method:

1. *Interaction effects* (substitutes and complements) between or among attributes are taken into account in the utility function, which assigns utility values to changes in the whole set of attributes, rather than attempting to "price" each attribute separately. In practice, however, attributes are sought that can be evaluated independently of each other.

2. The values of large *nonmarginal changes* in attribute levels are captured by the utility function formulation.

In addition, the multiattribute approach has several important advantages over holistic evaluation techniques that require that the proposed alternatives themselves be evaluated:

1. *Separation of expertise from preferences.* Individual preferences are

assessed over attributes that respondents are fairly familiar with and understand fairly well. The modelling required to predict or estimate the probable impact of proposed alternatives in terms of resulting changes in attribute levels can be done separately, using the best available techniques of impact assessment and statistical modelling. Respondents need not bear the burden of intuitive impact assessment, but need only express their preferences for results, as described through changes in attribute levels.

2. *Treatment of uncertainty.* Once multiattribute utility functions have been assessed, probabilistic uncertainty about the effects of proposed alternatives can be treated by computing expected utilities, using tools of analysis and modelling to supply the probabilities, and respondent utility functions to supply the utility values. This allows complicated probability distributions over utility values to be reduced to single utility numbers. Also, as just mentioned, it allows technical uncertainties to be modelled and assessed separately by experts, rather than forcing individuals to combine beliefs that they may be quite uncertain about with preferences that they are quite sure of.

Note: The substitution of expert beliefs for individual beliefs presupposes that experts are better informed than respondents (and impartial in their assessments). This assumption is quite plausible in some contexts – for example, in the assessment of the carcinogenic risks from new and unfamiliar chemicals. It is less plausible where individual behavior – for example, in the use of seatbelts – is a key source of uncertainty. In addition, the separation of preferences and beliefs is possible under the conventional SEU model (although the action taken is no longer prescribed for individuals based on their own subjective beliefs). However, this is not possible in some recent, more general theoretical models of "rational" behavior, for example, in models with state-dependent utilities (Karni, Schmeidler, and Vind 1983).

3. *Avoidance of strategic bias.* A major theoretical difficulty with benefits-assessment methods based on holistic evaluations is that they may tend to create incentives for respondents to misrepresent their true preferences (i.e., to "cheat") in an effort to manipulate the social decision mechanism to private advantage (Green and Laffont 1979). While there is some evidence that this is more of a problem in theory than in practice, and while there are theoretical reasons to believe that such problems can be largely overcome when there are sufficiently many respondents, it is plausible that incentives for strategic misrepresentation of preferences are effectively eliminated when respondents are asked about their preferences for attribute levels, rather than for concrete alternatives. This is especially true if the responses are to be used in evaluating a whole sequence of future proposals.

4. *Coverage of relevant attributes.* Empirical evidence suggests that real consumers in actual choice situations (involving choosing among competing

product brands) tend to choose on the basis of only a few attributes – often as few as two – even though the brands are described by many attributes. Different attributes are selected as the basis of choice on different trials by different consumers, however, so that the choice problem cannot be reduced to description in terms of only a few attributes (Horsky and Rao 1984). Such behavior is consistent with a random utility or stochastic choice model, in which alternatives are compared on the basis of the first few attributes that happen to come to mind. Different attributes may therefore be considered on different occasions, since the attributes that attract the respondent's attention on any particular trial are determined by an underlying probabilistic mechanism – perhaps to reduce the burden of information processing needed to make a choice.

If this model is accepted, then "true" individual preferences are defined only in terms of long-run averages in a hypothesized sequence of repeated trials, or in terms of population averages in a population of homogeneous respondents. The multiattribute approach is then seen as a short-cut method of arriving at these true preferences. By considering simultaneously all the attributes that an individual would eventually consider over the course of sufficiently many trials, the multiattribute utility model allows the long-term average responses to be predicted in a single pass. It takes into account all those factors that an individual might ideally wish to consider but that he might forget to consider in making any particular choice without the assistance of the model.

Two important caveats must be kept in mind in implementing the multiattribute approach. The first is that the predictive validity of the final model should be checked – for example, through use of a holdout sample during the preference elicitation and representation phase – by comparing the choices that it predicts with actual choices that have been carefully thought over (Currim and Sarin 1984). The model is intended to represent individuals' idealized (carefully considered, adequately informed) preferences, not to dictate them.

Secondly, it is of the utmost importance to make sure that important attributes (such as decision regret) that individuals would like to consider in their decision making are not omitted from or ignored by the model (Bell 1983). For example, the benefits of cleaner air or of a public health or safety regulation to an individual include, not only improvements in his own health and life expectancy, but also similar improvements for his friends and family. Omitting such altruistic externalities from the scope of the attributes for which preferences are elicited could give a very incomplete picture of the benefits to an individual from cleaner air and thus could lead to an underestimate of the benefits from such a measure. Checking predictive validity (as opposed to goodness-of-fit with past observed choices) of the final model helps to prevent such omissions.

Effective implementation of the multiattribute approach requires simple techniques for assessing the multiattribute utility functions $U(X_1, \ldots, X_n)$. These techniques form the core of multiattribute utility theory and of closely related techniques such as conjoint measurement of preferences (Keeney and Raiffa 1976; Krantz *et al.* 1971; Currim and Sarin 1984). The key steps in the assessment of multiattribute utility functions are reviewed below.

Selecting attributes and establishing independence conditions among them

The principal strategy in multiattribute utility assessment is one of divide and conquer. The first step is to separate attributes into mutually independent subsets such that the value of changes in attribute levels within one subset can be assessed without knowing anything about the changes in the levels of attributes in other subsets. Thus, all substitute and complement effects hold within, rather than between, subsets.

Important types of independence among attributes include the following:

1. *Mutual preferential independence* (MPI). The attributes $X_1, \ldots X_n$ are said to be mutually preferentially independent for a respondent if his preferences between any two distinct attribute bundles depend only on the attribute levels that differ between the two bundles and not on the attribute levels that are the same in both bundles (if any). In other words, his marginal rate of substitution between any two attributes does not depend on the levels of the other attributes.

2. *Utility independence* (UI). Attribute X_i is said to be utility independent of the remaining attributes for a respondent if his preference ordering for probability distributions over X_i does not depend on the levels of the remaining attributes.

3. *Difference independence* (Dyer and Sarin 1979). Attribute X_i is said to be difference independent (DI) of the remaining attributes if the preference difference between any two attribute bundles that differ only in the level of X_i does not depend on the common levels of the remaining attributes. Similarly, attribute X_i is *weakly difference independent* (WDI) of the remaining attributes if the *ordering* of preference differences among pairs of attribute bundles that differ only in the level of component X_i depends only on those levels of X_i and not on the common levels of the remaining attributes. (Here we are using X_i to denote both the *i*th attribute and its level. Also, recall that the concept of preference difference has previously been defined in terms of whether the change in going from one attribute bundle to another is more or less preferred to the change in going from a third attribute bundle to a fourth.)

4. *Additive independence*. Attributes $X_1, \ldots, X_n$ are said to be additively independent (AI) for a respondent if his preferences for a probability distribution over attribute levels depend only on the n marginal probability distributions of each attribute separately and not on their joint distribution.

In selecting attributes to describe and represent the probable effects of proposed alternatives, the analyst should keep these sorts of independence in mind. It is often possible to find many different sets of attributes, each of which suffices to describe the outcomes of proposed alternatives. The set that is selected should be *complete,* in the sense that descriptions of outcomes in terms of those attributes alone provide a sufficient basis for choosing among them, and *irredundant* (or minimal), in the sense that different attributes measure different constructs. Selecting attributes that satisfy the above sorts of independence will help meet the latter criterion, and will also greatly simplify subsequent analysis and interpretation. Moreover, practical experience suggests that it is usually possible to find such a set of attributes, given sufficient thought. Techniques for efficiently verifying various independence conditions can be found in the references, especially Keeney and Raiffa (1976).

Assessing component value and utility functions

Assuming that attributes are MPI, the next step is to assess a utility or value function over each attribute separately. If the consequences of each proposed alternative in terms of changes in attribute levels are known, then a value function may be assessed. If the consequences are very uncertain, then utility functions should be used. If the MPI and AI conditions discussed above hold, however, and if DI holds for some attribute, then under weak technical assumptions the value and utility functions will coincide, so that no distinction between them need be drawn (Dyer and Sarin 1979).

Even where these conditions do not hold, utility functions can often be obtained by simple deterministic transformations of the corresponding value functions. Under certain technical conditions, for example, $U(X)$ may be obtained from $V(X)$ by the transformation

$$U(X) = \frac{1 - \exp[-cV(X)]}{1 - \exp(-c)}$$

where c is a coefficient of relative risk aversion (positive for risk-averse respondents, negative for risk-seeking respondents; Currim and Sarin 1984 and references therein).

When $U(X)$ and $V(X)$ are identical, or when $U(X)$ can be simply obtained from $V(X)$, techniques for value function assessment, such as conjoint assessment techniques, can also be used to assess utility functions. It is good practice in such cases to use both sorts of techniques, to provide a check on the validity of the results.

The significance of the independence conditions discussed above, especially MPI and UI, is that they allow component value or utility functions to be assessed over each attribute separately, so that $U(X_1, \ldots, X_n)$ can be decomposed as a function of n component utility functions $U_1(X_1), \ldots, U_n(X_n)$. This substantially simplifies assessment of the overall (aggregate) utility function.

Techniques for assessing the single-attribute component utility functions are discussed in the references. Although there are a variety of potential techniques, ranging from direct assessment to lottery-based elicitations, simple methods such as direct assessment of value functions, or conjoint techniques such as the mid-value splitting technique (Keeney and Raiffa 1976) often seem to work best.

Direct assessment and lottery-based techniques have already been described for holistic utility functions that assess utilities for alternatives rather than attributes. Similar techniques can be used to assess value or utility functions over attributes. The difference is that utility functions for numerical attributes map one numerical scale (for the attribute level) into another (for the corresponding utility level), and thus can exploit properties such as monotonicity and continuity that were not available when discrete alternatives were being assessed.

The details of these techniques can be pursued through the references (especially Keeney and Raiffa 1976 and Fishburn 1967) and through recent papers in journals such as *Econometrica, Management Science,* and *Operations Research*. The principal conclusion is that component value or utility functions for each attribute can often be assessed fairly rapidly and reliably by first establishing a few qualitative properties (such as concavity or risk aversion) and then assessing a few points – generally no more than three – that indicate the overall shape of the function. After a component utility function has been estimated, it is verified and adjusted as necessary through a series of diagnostic questions that confirm its qualitative properties. A second opportunity for correction and verification is afforded when the whole utility function is checked for predictive validity, following aggregation of the component utility function.

Example 4: Applications to valuing health attributes

A number of applications of the multiattribute approach have been made in the difficult area of valuing human health. These studies have helped to illuminate the structure of human preferences both when the respondents' own health, or pain and suffering, are involved, and when some individuals (e.g., physicians or public policy makers) are required to make judgments involving the health states of others.

Pliskin and Beck (1976) presented an example in which an individual's value functions were assessed over five "difficult" attributes, including severity of heart disease, chronic bronchitis and emphysema, and diabetes. Their paper was intended primarily to illustrate the techniques and applicability of multiattribute evaluation and validation for difficult health-related attributes.

Similarly, Pliskin, Shepard, and Weinstein (1980) give an application of multiattribute utility (MAU) theory to the valuation of health status, with a specific application to anginal pain. This paper defines a class of MAU models for health status and life years and considers explicitly the tradeoffs between longevity and relief from anginal pain that would produce equal utility. In addition, relevant value and utility functions for several individuals were assessed empirically, and the SEU theoretical framework was used to derive conditions under which the use of expected "quality-adjusted life-years" in health status indices could be reconciled with the requirements of "rational" decision making.

More recently, Torrance, Boyle, and Horwood (1982) developed a sophisticated generic multiattribute approach to the valuation of health states, using four aggregate attributes (relating to physical function, role function, social-emotional function, and health problems) to classify overall health states. Both value and utility functions were assessed over each attribute for nearly 90 subjects, for use in a cost/social-utility analysis of neonatal intensive care. The precision and reliability of utility-assessment techniques for such health applications was examined, and the general usefulness and reliability of the approach for this context was demonstrated. This paper also considers several other aspects of the social-utility approach to evaluation of health effects and is recommended as a case study demonstrating the social utility approach to benefits assessment.

Additional applications of MAU assessments to valuation of human health include an examination of the roles of risk attitudes and time preference in determining individual utility functions for health effects distributed over time (Gafni and Torrance 1984). Krischer (1980) gives an annotated bibliography of over 100 recent (post-1970) articles and reports relating to single-attribute and multiattribute utility assessments, with applications to valuing health effects ranging from sore throat to cancer. This example of evaluating health effects will be continued below in discussing aggregation of component utility functions.

Summary and limitations of component utility assessment

The assessment of component utility functions for each attribute plays a role analogous to the assessment of implicit prices for attributes in the econometric approach. Namely, attribute levels are translated into utilities on a value or utility scale that is conventionally calibrated to run from 0 (least valued) to 100 (most valued). This is analogous to the translation of attribute levels into dollar values through implicit prices, except that the units are no longer dollars, and the translation may consist of something other than multiplication by a constant implicit price. In fact, the ith component utility function $U_i(X_i)$ will in general be nonlinear in the attribute level X_i (e.g.,

5. Theory of Regulatory Benefits Assessment

$U_i[X_i] = 1 - \exp[-cX_i]$ is one commonly used form). An individual's component utility functions therefore supply the information needed to evaluate nonmarginal changes in attribute levels – something that requires assessment of individual inverse demand curves if monetary willingness-to-pay measures of benefit are used.

Although the logical role of component utility functions is clear, two practical difficulties should be recognized:

1. A key implicit assumption is that component utility functions over attributes exist and can be objectively assessed through iterative questioning. Empirical evidence, however, suggests that utility functions for many attributes, such as number of lives saved per year by a proposed traffic safety measure, cannot realistically be modelled as mapping levels of the attribute into corresponding utility levels. Instead, the utility of changes in the attribute level depends on the status quo, or reference point, and the component utility function assessed over different final attribute levels will depend on the starting point, or current state. In effect, component utility functions are state-dependent, so that unique utility functions over attribute levels may not exist.

A precisely similar phenomenon arises when willingness-to-pay or direct assessments of monetary value are used to evaluate proposed changes. Utilities (or WTP values) for gains and losses are often inconsistent with utilities (or WTP values) for absolute attribute levels (Tversky 1979). In such cases, we suggest that utilities for changes in attribute levels relative to the status quo (if an unambiguous status quo point can be identified; see Chapter 4) are probably the most relevant guides to action. However, this makes the assessment of component utility functions dependent on the status quo.

2. Assessment of a utility function over even one attribute for even one individual requires some thought and time on the part of the respondent, and some time from a trained interviewer, unless printed questionnaires are used. (Although some assessments have been made using questionnaires, it is often necessary to have an interviewer present to help respondents understand the task, especially if more than one construct is represented by a single attribute; see e.g., Torrance, Boyle, and Horwood 1982.) Although direct assessment of value functions, for example, can proceed relatively quickly once the respondent understands what is required, the task of supplying preference data in this form is more difficult than if simple ordinal preferences or binary comparisons are required.

Aggregating component utilities

The final step in the multiattribute approach to utility assessment is to aggregate the component utilities assessed for the changes in individual attribute levels into an overall utility for the entire bundle of changes. One way to do this is suggested by analogy with the econometric approach. The utility or value of a change in several attributes might be found as the sum of the utilities or values of the change in each attribute separately. Clearly, however, such additive aggregation is justified only if the attributes are in some sense independent, so that the value of a change in several attributes is not greater or less than the sum of the values of the changes in each, as would occur if the changes were complements or substitutes, respectively. More exactly, we have the following results:

(1) If the attributes used to describe the effects of a proposed regulation are mutually preferentially independent (MPI), and if certain technical conditions hold, then the value of a bundle of changes in the levels of several attributes can be represented as the sum of the component values for the changes in the level of each attribute separately. Symbolically,

$$V(X_1, X_2, \ldots, X_n) = V_1(X_1) + \ldots + V_n(X_n) \qquad (1)$$

where $V_i(.)$ is the ith component value function and X_i is the level of the ith attribute (Krantz et al. 1971). (The analogous relation for *changes* in attribute levels and corresponding value levels obviously follows immediately.) Moreover, if the $V_i(.)$ are measurable value functions (so that differences in values represent differences in preference intensities; see definitions above), and if difference independence holds for any attribute, then the aggregate value function given by Eq. (1) is also measurable. Value functions that are not measurable – that is, for which value differences have no empirical significance – are sometimes referred to as conjoint models.

If the effects of a proposal are very uncertain, so that component utility functions are used to represent preferences for probability distributions over attribute levels, then a result for utility functions that is very similar to the above result for value functions is useful:

(2) If the attributes $X_1, \ldots X_n$ are additively independent and some other mild technical conditions are met, then the aggregate utility function for multiple attributes is just the sum of the component utility functions for each attribute separately. Symbolically,

$$U(X_1, \ldots, X_n) = U_1(X_1) + \ldots + U_n(X_n) \qquad (2)$$

Thus, the change in utility from a change in the levels of several attributes simultaneously is the sum of the changes in component utilities from changes in the level of each attribute separately.

These results are very useful in cases where the required independence conditions can be established. However, there are other forms of aggregation than addition, and these may be required if a set of attributes for which these independence conditions hold cannot be found.

(3) If attributes $X_1, \ldots, X_n$ are MPI, and WDI holds for at least one attribute, then the value function for a bundle of attributes is obtained by multiplicative aggregation of the component value functions, according to the following aggregation formula:

$$V(X_1, \ldots, X_n) = \frac{\Pi_i[1 + KK_iV_i(X_i)] - 1}{K} \tag{3}$$

if $K_1 + \ldots + K_n \neq 1$; and by the additive aggregation formula

$$V(X_1, \ldots, X_n) = K_1V_1(X_1) + \ldots + K_nV_n(X_n)$$

if $K_1 + \ldots + K_n = 1$. Here Π_i denotes the product over all components (i.e., from $i = 1$ to n), and all the value functions, both the components $V_i(.)$ and the aggregate $V(.)$ are assumed to have been scaled to run from 0 to 100. The scaling constants K and K_i are chosen accordingly.

(4) If $X_1, \ldots, X_n$ are MPI, and if utility independence (UI) holds for at least one attribute, then utilities are aggregated multiplicatively; that is, the above result Eq.(3) for value functions holds for utility functions, with $U(.)$ and $U_i(.)$ replacing $V(.)$ and $V_i(.)$.

An important feature of the multiplicative aggregation rule is that it allows for possible substitute or complement effects among attributes. Specifically, if the scaling constant K satisfies $K < 0$, then the attributes are substitutes and the respondent is said to display multivariate risk aversion. Conversely, if $K > 0$, the attributes are complements, and the respondent is multivariate risk-seeking. In either case, multiplicative aggregation is required. Only if $K = 0$ (multivariate risk neutrality, no preference interaction among attributes) is the additive aggregation rule justified (see e.g., Torrance, Boyle, and Horwood 1982).

The basic reference for these results, and for details on how to calculate the scaling constants K and K_i, $i = 1, 2, \ldots, n$, is Keeney and Raiffa (1976), which also gives advice on what to do if not even these weaker independence conditions can be established. (The most common solution is to rethink the choice of attributes used to describe the problem, e.g., by reaggregating more primitive attributes into different high-level attributes.) The extension to measurable value functions is due to Dyer and Sarin (1979). As previously mentioned, Torrance, Boyle, and Horwood (1982) provide an excellent case study on the social-utility approach to the assessment of health states and discuss the many practical details of implementation.

Example 4 (cont'd): Additive versus multiplicative aggregation

In many, if not most, applications of multiattribute preference assessment, the additive aggregation form of Eq.(1) for values or utilities has provided an adequate representation of individual preferences, and use of the more complicated multiplicative form has not been justified by a significant increase in predictive validity. For example, Tables 5–1 and 5–2, from Currim and Sarin (1984), show the predictive validities for these forms and for various assessment strategies (e.g., statistical vs. algebraic) in a job

Table 5–1. Predictive accuracy of statistical and algebraic models under certainty

Model	First validation sample		Second validation sample	
	Proportion of correct predictions on holdout sample	Kendall's tau	Proportion of first rank correctly predicted	Proportion of all ranks correctly predicted
Statistical				
Additive Conjoint				
–Partworth	0.80	0.72	0.74	0.66
–Linear	0.63			
Additive Measurable Value				
–Partworth	0.81	0.66	0.60	0.59
–Linear	0.64			
Additive Utility				
–Partworth	0.75	0.67	0.69	0.63
–Linear	0.63			
Naive	0.47			
Algebraic				
Additive Conjoint	0.66	0.54	0.55	0.56
Measurable Value				
–Additive	0.65	0.21	0.46	0.30
–Multiplicative	0.65			
Utility				
–Additive	0.59	0.24	0.51	0.36
–Multiplicative	0.68			
Linear Programming				
Additive Measurable Value	0.63			
Additive Utility	0.57			

Source: Currim and Sarin (1984)

choice application. It is clear that the best results are obtained from the additive aggregation form.

An important exception to this rule appears to be in the assessment of multiattribute utility functions for health effects. Torrance, Boyle, and Horwood (1982), in the case study previously mentioned, found strong evidence that the additive model is inappropriate, and that health state attributes tend to be complementary rather than preferentially indepen-

5. Theory of Regulatory Benefits Assessment

Table 5–2. Predictive accuracy of statistical and algebraic models under certainty

Pearson correlations between actual stated p-values for four options and predicted values of holdout

Models	Pearson correlation
Statistical	
Additive Conjoint	0.62
Additive Measurable Value	0.64
Additive Utility	0.84
Algebraic	
Additive Conjoint	0.16
Additive Measurable Value	0.08
Additive Utility	0.14

Source: Currim and Sarin (1984)

dent; they therefore recommend multiplicative aggregation. Similar results have been reported by other investigators using different health status attributes.

Example 5: Valuing risk reductions

The multiattribute utility theory outlined above for individual preferences can be used not only to quantify (idealized) individual preferences among alternatives in specific cases, but also to derive some general theoretical insights that are useful in the assessment of health and safety benefits. For example, Weinstein, Shepard, and Pliskin (1980) use a simple two-attribute model to study the tradeoffs between dollars or asset position (the first attribute) and risk to life (the second). Under the assumptions that (a) life is preferred to death for any asset position; (b) the marginal value of an increase in assets is greater in life than in legacy; (c) a lottery giving a 50 percent chance at death with no legacy and a 50 percent chance at life with asset position x is preferred to death with legacy x, for any x; and (d) individuals are risk-averse in assets, Weinstein *et al.* show that

1. The dollar value of a risk reduction, as measured by WTA or WTP, depends not only on the size of the risk reduction but also on the size of the baseline risk being reduced, where the risks being considered are risks to life. Specifically, a risk reduction of a given magnitude is more valuable in terms of both buying and selling prices when it is made in a larger initial risk level. For example, a risk reduction from 0.4 to 0.3 would be more valuable, under the above assumptions, than a risk reduction from 0.3 to 0.2.
2. The selling price per expected life saved (i.e., the selling-price value of

a statistical life saved to its owner, in modern terminology) is greater for larger risks than for smaller ones. In particular, the WTA value for reducing a risk from 0.03 to 0.02 is less than one tenth of the value of reducing it from 0.3 to 0.2. However, the WTP, or buying price, per expected life saved may be greater in the former than in the latter case, in contradiction to the WTA price.

3. Societal WTP, defined as the sum of individual WTP values, will tend to be greater for a health or safety regulation that saves one life with certainty than for a regulation that saves ten lives with probability 0.1 each.

Perhaps the most important overall conclusion from this research is that a unique dollar value per expected life saved, even as valued by the owner of the life, may not exist. Instead, this value will depend on the initial level of the risk being reduced. Thus, for small risks ranging over several orders of magnitude, one would expect to find a range of dollar values for statistical lives saved (i.e., dollars per expected life saved.) Therefore, it would be senseless to search for a universally applicable dollar value per expected life saved, even for small risk reductions.

This example illustrates the types of qualitative implications for health and safety measures that can be drawn from multiattribute utility theory. It has several noteworthy features. First, the data requirements needed to draw useful qualitative conclusions are relatively mild, having to do primarily with the qualitive properties of individual risk attitudes. Second, it is clear that some relatively challenging policy questions involving health and safety issues can at least start to be addressed through the use of simple analytic models having mild enough requirements so that empirical preference patterns might reasonably be supposed to follow the theoretical predictions of the models fairly well.

Finally, this example illustrates one common way of bridging the apparent gap between utility-based and money-based measures of individual preference. Namely, a multiattribute utility function having monetary position as one attribute is assessed and used to define indifference curves or tradeoffs between money and traditionally hard-to-price attributes such as risks to life. The resulting tradeoff rates establish dollar equivalents for the hard-to-price attributes with respect to the underlying multiattribute utility model. (Of course, unlike true market prices, such WTP or WTA figures in general apply only to the individual whose preferences are being assessed: there is no necessary equilibrating adjustment mechanism to equate the tradeoff rates for different individuals.) A more general discussion of this strategy for converting from multiattribute utility to dollar representations of preferences may be found in Keeney and Raiffa (1976) under the name "pricing out." In practice, the technique is only useful when, as in the above example, there are only a very few attributes and certain simplifying qualitative assumptions about preference tradeoffs can be made.

5. Theory of Regulatory Benefits Assessment

Summary of component aggregation

To assess a respondent's value or utility from a change in the levels of several independent attributes, it is necessary to aggregate the values or utilities from changes in each component attribute. This is similar to the evaluation of a market bundle by summation of the values (price times quantity) of each component good in it. However, aggregation of component value or utility functions need not be additive. In fact, if attributes are substitutes or complements, then the additive form is inappropriate and a multiplicative aggregation rule should be used instead, assuming that the required independence conditions (MPI and UI or WDI) can be established. This clearly destroys the analogy with market prices for continuously divisible goods, where additive aggregation is always appropriate for marginal changes in the attribute levels.

If the independence conditions needed to justify additive or multiplicative aggregation cannot be established, then more complicated forms of aggregation are, in principle, required. In practice, however, it is usually preferable to restructure the top-level attributes used to describe the effects of proposed alternatives so that independence conditions are obtained. Experience suggests that this can usually be done.

Limitations and data requirements

After a multiattribute value or utility function has been assessed using either Eq. (1) or (3), it should be checked for predictive validity to make sure that all important constructs that help determine preferences have been adequately represented in the selected attribute set, and to ensure that the final aggregate function does indeed properly represent preferences. There is a conceptual difficulty with this diagnostic checking, however. Suppose that the preferences predicted by the multiattribute utility (MAU) model differ from the respondent's holistically assessed preferences. What does this imply about the validity of the MAU model? The purpose of using a MAU model to begin with is that holistic assessments are unreliable when alternatives differ on many independent attributes. (A well-known psychological rule of thumb for the value of "many" in situations like this is 7 ± 2.) The MAU approach provides a systematic methodology for decomposing difficult preference judgments into simple components and then aggregating the results to obtain an "idealized," constructed, preference pattern. To argue that this idealized preference pattern is inadequate because it fails to agree with holistic preferences misses the point of the MAU decomposition. On the other hand, the MAU model requires empirical checking.

A partial resolution of this dilemma is that diagnostic checking should focus on a search for systematic discrepancies between predicted and actual holistic preferences, or residuals that are sufficiently large and disorderly as to suggest that one or more explanatory variables have been omitted. In addition, the holistic preference judgments used to check the model should be carefully considered and perhaps reflect the average of several test-retest trials. As yet, however, there is no established methodology for implementing these suggestions. In particular, there is no well-developed error theory for supporting statistical tests of the adequacy of multiattribute models, given that respondents may be unsure about or mistaken about their own preferences on any particular trial.

Perhaps the best available substitute comes from logit and other econometric probabilistic choice models that assume that a "true" underlying representation for preferences of the form of Eq. (1), for example, exists, and then estimate its components and scaling constants so as to make the resulting model as consistent as possible with observed holistic response data. As previously mentioned, this statistical approach to the derivation of a MAU model provides an alternative to the assessment and aggregation of component utility functions. And Tables 5-1 and 5-2 suggest that this approach works well in at least some contexts. However, it is in general not suitable for estimation of multiplicative, as opposed to additive, models, since an extremely large set of response data is needed to adequately fit a model of the form of Eq. (3).

A final limitation of the MAU method is that it requires a considerable amount of preference data from respondents. Not only must component value or utility functions be estimated, as previously discussed, but the scaling constants used in the aggregation rule must also be estimated. (See Keeney and Raiffa 1976 and the case studies cited in Example 4 for details on how to estimate these scaling constants.) Ideally, cardinal preference data, for example, expressed as numbers on a measurable value scale, should be collected, so that component functions can be estimated easily and reliably. In addition, the multiplicative aggregation formula must be checked.

There are, however, several ways in which these data requirements can be reduced. For example, attribute weights can be estimated from ordinal preference data if it is assumed that preferences for alternative attribute combinations can be expressed in terms of the "distances" of these combinations from some "ideal" combination (Horsky and Rao 1984). More generally, cardinal MAU models can be estimated from ordinal preference data using MANOVA, logit models, or other similar statistical and econometric techniques. And preferential independence conditions for each individual can be tentatively assumed to hold, and then checked for consistency with the aggregate data (see Torrance, Boyle, and Horwood 1982 for details). Such simplifications bring MAU within the realm of the practical but do not eliminate the need for substantial data collection when large (nonmarginal) changes in attributes that may be substitutes or complements are to be evaluated.

Aggregating Welfare Changes across Individuals

The techniques given in the preceding section are directed toward the assessment of the changes in the welfare levels of individuals resulting from changes (or probable changes) in the levels of attributes. Individual welfare is taken to be synonymous with individual utility, as determined from (possibly idealized) individual preferences, and as measured by the holistic or MAU techniques described above. In this section we will assume that predicted changes in individual welfare from a proposed regulation have been assessed in a properly constructed sample of respondents, and will consider how to aggregate these changes in individual welfare levels into an overall measure of the gain in social welfare from the proposed regulation.

Aggregation of individual welfare changes across individuals necessarily involves one of the most controversial problems in applied welfare economics: comparison of the utility gains or losses to different individuals. In much of traditional welfare economics, this problem is avoided by focusing solely on questions of (Pareto) efficiency. For example, if the gainers from a proposed regulation could more than compensate the losers, and if the losers could not successfully "bribe" the gainers to forego the regulation, then it might be decreed that the net change in social utility or welfare from implementing the regulation would be considered positive, without worrying further about measuring or defining the magnitude of the increase (see Chapter 2).

In regulatory benefits assessment, however, it is conventional to try to measure benefits numerically (and cardinally; see Chapter 4), rather than in terms of the partial orderings provided by the Pareto-efficiency criterion or by various compensation tests (e.g., the Hicks-Kaldor test or potential Pareto improvement test). The goal is not just to be able to compare various alternatives against the status quo but to be able to compare the alternatives themselves using the anticipated magnitude of each one's contribution to social welfare. This more ambitious level of measurement, while not strictly required for effective public decision making, is often seen as the goal of quantitative benefits assessment. It requires individual welfare levels or utilities to be aggregated into a corresponding level of social welfare or utility. The function that accomplishes this aggregation is referred to as a *social welfare function*.

Symbolically the problem of aggregating utilities across individuals can sometimes be posed as follows. Let $U_i(X_1, \ldots, X_n)$ denote the level of the ith individual's utility level when the levels of attributes $1, \ldots, n$ are $X_1, \ldots, X_n$, respectively. (Here X_1 might measure the level at which some public good is provided: X_1 could be an air quality index.) Given these individual utility functions, we would like to define a social utility function $U(X_1, \ldots, X_n)$ mapping attribute levels into a measure of total social welfare. Moreover, the social utility function might plausibly be required to depend on the objective attribute levels only through the individual utility functions, so that there are no "merit goods." In this case, the social utility function can be written in the form $U(U_1, \ldots, U_n)$, where U_i still represents the utility level of the ith individual, and the attribute levels on which individual utilities depend are now left implicit. Such a function may be called a (cardinal) social welfare function.

Cardinal versus ordinal measurement

The requirement that $U(.)$ be a cardinal function is important. A less ambitious level of measurement would have the welfare aggregation procedure take individual preference orderings of proposed alternatives as input and attempt to produce a social ordering of the alternatives as output. This would clearly require less information than the social utility function approach. The difficulty with this scheme, however, is that it extracts too little information about individual preferences. A celebrated theorem due to Arrow (1963) shows that, in general, no ordinal aggregation procedure which maps individual preference orderings into a social preference ordering can simultaneously satisfy certain apparently reasonable principles of aggregation and be successfully applied to all possible sets of individual preference orderings. If individual preference orderings are sufficiently different, any ordinal aggregation scheme will fail (see, e.g., Luce and Raiffa 1957). Subsequent work has shown that consistent ordinal aggregation is possible if individual preferences exhibit limited agreement (Sen 1970) or satisfy other restrictions, and additional results and extensions continue to appear in the collective choice literature. But the inadequacy of ordinal measurement as a general basis for aggregating arbitrary individual preferences remains. We shall therefore concentrate on aggregation of utility functions.

Basic strategy

Given a social utility function $U(.)$, the change in social welfare from a proposed regulation is found as follows:

1. For each respondent i, the probable change in attribute levels from the proposed regulation and for that individual is predicted. (The change in an attribute's level from a regulation may be different for different individuals – as in the case of air quality improvements from an environmental regulation, for example, which affects different individuals differently, depending on where they live.)

2. The change in each affected individual's welfare as a result of the changes in attribute levels is deduced from his utility function. Specifically, if the old attribute levels were $X_1, \ldots, X_n$, and if the new ones produced by the regulation are $X_1^*, \ldots, X_n^*$, then the change in this welfare is $U_i(X_1^*, \ldots, X_n^*) - U_i(X_1, \ldots, X_n)$, where $U_i(.)$ denotes the utility function for individual i. We will abbreviate this change by letting U_i denote the utility

5. Theory of Regulatory Benefits Assessment

level for individual i without the regulation, and by letting U_i^* denote his utility level after the regulation; thus, $U_i^* - U_i$ is the predicted change in the ith individual's utility from the proposed regulation. If individual welfare were measured in monetary terms, and in the absence of income effects (so that individual utility could be assumed to be linear in money), then $U_i^* - U_i$ could be interpreted as the maximum amount that an individual would be willing to pay to obtain the regulation (if it is positive) or as the minimum amount that he would have to be paid (if it is negative) to make him as well off with the regulation and the payment as without the regulation and without the payment. In nonmonetary terms, it is the utility, to him, of the regulation.

3. The corresponding net change in total social utility or welfare is then $U(U_1^*, \ldots, U_n^*) - U(U_1, \ldots, U_n)$, assuming that the social utility function $U(.)$ is given.

If the net change in social utility from a regulation is negative, then the regulation would presumably not be implemented. Choice among alternatives can be made on the basis of maximizing the net gain in social welfare, as required by E.O. 12291.

Deriving the social utility function

The question of how to choose the aggregation procedure, or social utility function $U(U_1, \ldots, U_n)$, can be resolved in a number of different ways, depending on what principles or axioms are used as a basis for aggregation. The general strategy of the *axiomatic approach* is to require the aggregation procedure to satisfy certain abstract normative properties and then to derive the form of the aggregation rule from these properties. Particular implementations of this method include coherence, ethical, and efficiency approaches.

> COHERENCE APPROACHES: If it is assumed that $U(.)$ should satisfy the same axioms of rationality as individual utility functions, and in this sense be coherent, and if the UIIGI (unanimous indifference implies group indifference) principle of Chapter 4 is accepted, then it can be deduced that *the social utility function is just a weighted sum of the individual utility functions* (or a weighted average of them, depending on the choice of scale). The weights may, of course, all be equal. This result is due to Harsanyi (1952); similar results, using slightly weaker definitions of coherence have been obtained by other researchers.

A limitation of the coherence approach is that it cannot in general be extended to Bayesian group decision making with separate aggregation of group beliefs and group preferences (Raiffa 1968); also it is not clear why a group should be expected to behave coherently if its members disagree. Further exploration of the coherence concept leads to the literature on "syndicates"; see, e.g., Amershi and Stoeckenius (1983) and references therein.

ETHICAL APPROACHES: The conclusion that social utility can be represented as a weighted sum of individual utilities is also reached if we accept the following ethical axioms or principles: *Principle 1:* Social utilities should depend only on individual utilities (which has been assumed throughout our discussion of social welfare functions). Thus, there are no merit goods that are intrinsically desirable, apart from people's preferences. *Principle 2*: If everyone has the same utility function, then the aggregate social utility function should be this function (or should be strategically equivalent to it, meaning that it is a scaled version of the common utility function, and hence leads to the same decisions). *Principle 3*: If everyone in society except for one person is indifferent between two proposed alternatives, then society's preferences between the two alternatives should just be that individual's preferences (i.e., the one individual who cares is decisive for society in this extreme situation). If these three principles are accepted, then it can again be deduced that $U(U_1, \ldots, U_n) = K_1U_1 + \ldots + K_nU_n$, where the K_i are equity weights to be determined.

This result is due to Keeney and Kirkwood (1975). In the same paper they note that the UIIGI axiom and the resulting additive aggregation scheme may not be desirable on ethical grounds. We can paraphrase and extend their argument by noting that additive aggregation draws no distinction between the following situations:

a. Half of the individuals in society receive their most-preferred outcomes and the other receive nothing (which can be assumed to be their least-preferred outcome). It is known in advance who falls in which half. This could correspond to a distribution of benefits to some people (e.g., from a health or safety regulation with benefits only to a certain age group or other sensitive subpopulation) that is paid for by everyone. Those who benefit gain from the regulation and financing package, while those who do not benefit lose from the combined package.

b. Each individual independently has a 50 percent chance of receiving his most-preferred outcome (utility = 100), and a 50 percent chance of receiving nothing (utility = 0). This might describe a safety regulation such as mandatory airbags that has a high chance of saving some lives but where those whose lives will be saved cannot be distinguished in advance. (This is also the sort of problem in which state-dependent utilities arise: each individual may prefer not to pay for airbags, unless his life is saved by one; see Karni, Schmeidler, and Vind 1983.)

c. There is a 50 percent chance that everyone will obtain his most-preferred outcome, and a 50 percent chance that no one will gain anything. This could apply to a safety regulation that is similar to the mandatory

5. Theory of Regulatory Benefits Assessment

airbags, but which affects a catastrophic risk (e.g., safety precautions on a threat that could potentially affect everyone) rather than distributed, individual risks.

d. Exactly half the people in society will gain their most-preferred outcomes, and the other half will gain nothing, but there is no way of telling in advance who will gain and who will lose.

It is assumed for simplicity that in each situation each individual's utility depends only on his own consumption of regulatory benefits, so that individual utility is not defined over societal distributions of consumption in such a way as to prohibit any of these situations.

Since the expected utility to each individual is the same in situations (b), (c), and (d), namely 50, the UIIGI principle would require that all three be indifferent from a societal perspective. Yet it is certainly conceivable that a public decision maker might prefer situation (b) to the others, on the grounds that it gives everyone a fair chance of gaining, while virtually guaranteeing (if there are many individuals in society) that someone will in fact gain. Moreover, if additive aggregation with equal weights for all individuals is used, all four situations will give the same social utility, even though considerations of prior and posterior equity might seem to make some of these situations less desirable than others.

In summary, additive aggregation and the UIIGI principle suffer from the defect that correlations in the pattern of uncertain benefits distributed to different individuals are totally ignored. Yet equity considerations may require such correlations to be taken into account.

One way to overcome this deficiency is to weaken the principles that the aggregation formula is required to satisfy. For example, instead of the unanimity condition that the social utility function should be a (possibly scaled up) version of the common individual utility function if a common utility function exists, we might require only the following *Principle 2'*: The social preference between any two alternatives should depend only on the preferences of those individuals who are not indifferent between the two alternatives (i.e., only the preferences count of those who care). If this is substituted for the unanimity condition (Principle 2 above), then a weaker form of aggregation rule is obtained. Namely, the only aggregation formula (social welfare or utility function) for individual utilities that satisfies ethical Principles 1, 2', and 3 is the multiplicative aggregation formula (3) introduced in the discussion of attribute aggregation (Keeney and Kirkwood 1975). In effect, each individual's utility is treated as one attribute in describing overall social welfare. This reduces to the additive aggregation formula only in the special case where the scaling constant $K = 0$. If all four principles, 1, 2, 2', and 3, are accepted, then of course the additive aggregation form is again required.

Using a multiplicative instead of an additive aggregation form for the social utility function introduces undesirable complexity into the aggregation process and does not necessarily succeed in distinguishing between situations c and d above in terms of differences in social utility.

Rather than giving up any of the Principles 1 to 3 and relying on a multiplicative aggregation rule, therefore, a more satisfactory approach to incorporating equity concerns into the social utility function may be to include the distribution of physical or economic benefits from a proposed measure among the attributes used to describe proposals. Individual concerns about equity can thus be introduced directly into the assessment of individual utilities for proposed measures and so enter into the determination of social utilities.

EFFICIENCY APPROACHES: A third approach that also leads to the additive aggregation form considers the Pareto-efficiency of the social utility function. Specifically, suppose that the social utility function is a continuous function (so that small changes in individual utilities correspond to small changes in social utility), and that it is required to produce social decisions – defined as those that are expected to maximize the gains in social utility – which are Pareto-efficient, no matter what the individual utility functions being aggregated are. Then the social utility function must be a weighted sum of individual utility functions (or a rescaled version of such a sum). Loosely speaking, the additive social utility function is the only one that guarantees Pareto-efficient social decisions (Kirkwood 1979).

In summary, different lines of reasoning based on coherence, ethical, and efficiency arguments all lead to the conclusion that the social utility function should be a weighted sum or average (if the sum is scaled down by $1/n$) of the individual utility functions. While this form of aggregation may ignore some equity concerns unless they are explicitly included among the attributes used to describe alternatives to individuals, no other aggregation rule is as simple or has such a strong theoretical basis. We shall therefore assume henceforth that *aggregation of welfare changes across individuals is accomplished by weighted summation*. Note that when social utility is a weighted sum of individual utilities, the change in social utility from a proposed measure is just the corresponding weighted sum of the changes in individual utilities – a convenient relation that does not hold for other forms of aggregation.

There are many other approaches to the aggregation problem that extend or complement those mentioned above. For example, Fishburn (1984) summarizes six proposed equity axioms for evaluating the distribution of public risks from alternative proposals and identifies subsets of these axioms that are mutually consistent. Chew (1983) considers a class of problems in which distributions (e.g., of benefits across individuals) are to be assessed for equity and reduced to numerical equivalents and concludes that the most general way of accomplishing this reduction, under certain coherence and

5. Theory of Regulatory Benefits Assessment

efficiency assumptions, is by use of a "quasilinear mean." This result is applied to synthesize problems in the measurement of income inequality with problems in the choice of a social welfare function.

Choosing the equity weights

Assuming that aggregation of benefits across individuals is to take place through weighted summation, where both individual and social benefits are measured in terms of changes in utility, the problem arises of choosing the equity weights $K_1, \ldots, K_n$ for use in the aggregation. One possibility is to use "relative needs" to determine these weights (Brock 1980). Relative contribution or willingness to bear a priori risks in exchange for potential a posteriori benefits should also perhaps be considered. However, such approaches have not been fully worked out even in theory (except for the special case of risk-sharing "syndicates"; see, e.g., Raiffa 1968), and are certainly not ready for practical application. Bodily (1979) has provided a theoretical procedure for explicitly calculating equity weights as the equilibrium outcome in a process of successive rounds of "delegation." This approach, however, would only be practical in small groups.

In practice, the usual approach is to assume that *all individuals are weighted equally*. Of course, this is only useful if it is known how individual utilities are to be compared, since the equity weights can really be thought of as the products of (1) scaling constants used to put utilities for different people in the same units; and (2) importance weights used to determine how heavily the welfares of different people are to be counted in assessing total social welfare.

One way of putting the utilities of different people on the same scale is to use endpoints (such as "normal healthy life" and "early and painful death" in a health effects example) which everyone agrees are at least as good and at least as bad as any of the alternatives being considered, and then to *assume that the utility difference between these two endpoints is the same for everyone*. This is the approach taken in the health effects case study of Torrance, Boyle, and Horwood (1982), for example (Example 4). Once individual utility scales have been coordinated in this way, social utility can be defined as the arithmetic average of individual utilities.

Discussion and comparison with monetary aggregation methods

The social utility function approach to aggregating individual utilities embodies an ethical assumption that individual utilities can be compared, at least if cardinal measurement is used and common endpoints are used to calibrate the individual utility scales. The need for such value judgments in the choice of equity weights is sometimes criticized as a weakness of the social utility approach, especially when it is seen as putting social value judgments into the hands of the analyst.

The usual response to this criticism is that any approach to aggregation of individual benefits requires value judgments, either implicitly or explicitly. The assumption that the utility difference between endpoints on an outcome scale is the same for different individuals requires exactly the same sort of value judgment as the assumption that a dollar is to be considered equally valuable (in terms of net social welfare) no matter who receives it – and is, perhaps, more defensible. Indeed, it can be shown that the Pareto-efficient level of a public good is independent of income distribution only under extreme and implausible assumptions about individual utility functions (essentially, that they are quasilinear in income; see Bergstrom and Cornes 1983). Thus, the logic of additive aggregation of utilities is similar to the logic behind traditional monetary aggregate benefit measures, such as consumer surplus, but is if anything somewhat easier to interpret and justify.

The objection that choice of equity weights – for example, the use of equal weights for different individuals – should not be left up to the analyst can be met in principle through the use of sensitivity analysis. Namely, the weights assigned to different stakeholder groups (i.e., to the individuals in them; the two concepts coincide under additive aggregation) can be varied to determine how sensitive the final choice among alternatives is to the choice of aggregation weights. As previously mentioned, it is generally desirable to present the results of a benefits assessment in sensitivity analysis form, so that the user of the assessment can see the effects of using different value judgments, including his own. Additive aggregation through equity weights makes possible sensitivity analyses that are particularly easy to conduct and interpret, and allows especially informative and effective presentations of the results to decision makers (see von Winterfeldt and Edwards, in press).

Time, Risk, and Uncertainty in the Social Utility Model

We will now concentrate on the problem of deriving a social utility function for probable time streams of consequences. To use this utility function in a sequential choice process for optimizing social policy, it is necessary to use dynamic programming and principles of rational sequential choice (see, e.g., Whittle 1982 and Kreps and Porteus 1978, respectively) to construct a sequence of actions and research that balances the costs of deferring decisions against the costs of making incorrect decisions, and that exploits the opportunities for information collection. The social utility function is used in such dynamic policy choice problems both to quantify the value of collecting additional information and to compute the expected utilities of alternative next steps in the policy with respect to the current information base, when the expected value (or rather utility) of collecting additional information falls below the expected loss in utility from deferring

5. Theory of Regulatory Benefits Assessment

decision any further. As in the case of static choice, however, techniques for constructing an effective decision procedure go well beyond the scope of benefits assessment per se and are not considered further in this study.

Individual Value and Utility Functions for Time Streams

Three general approaches to dealing with time, risk, and uncertainty – the decentralized, market, and centralized approaches – have been discussed already, and the centralized approach has been suggested as being the most effective for benefits assessment. The first step in the centralized strategy as applied to utilities is to obtain individual value and utility functions for time streams of attribute levels consumed or experienced by the respondents. This can easily be done, in many circumstances, by extending the theory of multiattribute assessment to attributes in different periods.

For example, suppose that attribute levels in different periods are mutually preferentially independent (MPI), so that the utility of consumption in one period, for example, does not affect the utility of consumption in other periods. (For this to be reasonable, periods have to be defined as appropriately long intervals.) Suppose also that a respondent's preference ordering of any two time streams that differ (diverge) only after the first period is the same as his ordering of the two substreams starting with the second period (as he would evaluate them from the start of the second period, assuming that his preferences are stationary, so that they depend only on the future sequence of attribute levels and not otherwise on the period). Then under mild additional technical assumptions, it follows that the value of a time sequence $X(1), X(2), \ldots$, where $X(t)$ denotes the bundle of attribute levels in period t, can be expressed in the form $v[X(1)] + av[X(2)] + a^2v[X(3)] + \ldots$, where $v(.)$ is the single-period value function, and a is a discount factor representing the individual's time preference rate. (See Keeney and Raiffa 1976: Chapter 9; this result is due to T. Koopmans.)

The derivation of utility functions for uncertain time streams is somewhat more complicated. We refer the reader to Bell (1977) for an excellent and detailed technical summary and discussion. The principal results are that one-period or two-period individual utility functions can be aggregated over time (assuming a fixed time horizon) using additive or multiplicative aggregation formulae (or, as Bell shows, a ratio of multiplicatively aggregated one-period and two-period utility functions, or a difference of additively aggregated one-period and two-period utility functions), assuming that appropriate independence conditions can be established. Bell provides a simple test for determining whether additive or multiplicative aggregation should be used, and summarizes an application to a forest-pest problem with three time streams of attributes: profit to the lumber industry, employment within the industry, and recreational potential of the forest.

Planning horizons and mortality

A special difficulty arises in the derivation of utility functions for probable time streams of consequences when the mortality of the respondent is considered. The effect of risks to life and of uncertainty about life span is to make the time horizon over which utilities are assessed a random variable.

The value of small risk reductions (corresponding to a probabilistic forward shift of the individual's horizon) and tradeoff rates between consumption and expected remaining years of life has been studied in the recent literature (Howard 1984; Shepard and Zeckhauser 1984) in the context of deriving dollar-value "equivalents" for small risk reductions. However, these approaches make use of quite specialized theoretical models. There is as yet no applicable general theory for evaluating time streams in which survival probability may itself be one of the attributes in each period, as well as determining the probable number of periods.

Possible simplifications: planning horizons and average costs

In some applications, intertemporal benefits assessment can be simplified by the use of a finite *planning horizon*. This is most appropriate when the consequences of each current alternative are known and deterministic in terms of (a) immediate resulting changes in attribute levels, (b) changes in the state of the world (e.g., the amount of a scarce resource left after the current consumption decision), and (c) changes in the set of future opportunities for decision making; and when the rate a at which future benefits are discounted in the intertemporal value function, as discussed above, is not too small (Bean and Smith 1984). Another possible simplification that may be appropriate if the time stream of benefits is essentially stationary (although possibly with random fluctuations about a long-run average) is to use the average (expected) benefit per period as a criterion, thus bypassing the need for discounting. Such simplifications do not, however, address or overcome the fundamental difficulties associated with uncertain life spans and irreversible transitions in the state of health or well-being of individuals.

Requirements for intertemporal utility functions

The fundamental problems in constructing useful intertemporal utility functions can be summarized as follows:

CONSISTENCY: An idealized intertemporal utility function should give the same ranking of alternatives to be chosen from at a given date, no matter when the assessment is made (assuming that information is held fixed). That is, in the absence of new information one's decision should not change as the decision date approaches, nor should one deliberately make decisions that he will later regret. Short-run and long-run preferences, given an information set, should ideally coincide (Kreps and Porteus 1978).

IRREVERSIBLE CHANGES: The utility to a respondent of a particular flow of consequences may depend on his state when he receives it. An annuity of

$100,000 per year may be worth less to someone who is crippled or in jail than it would be to him if he were free and healthy. Hence, an intertemporal utility function may have to incorporate *state-dependent* utilities, allowing for the possibility that the utility of a time stream of consequences will depend on the (currently uncertain) future state of its recipient (Karni, Schmeidler, and Vind 1983). Similarly, an individual may be uncertain about his own future preferences, and current decisions (and hence intertemporal utility functions) should allow for such uncertainty.

ANXIETY: An additional consideration in the construction of intertemporal utility functions is that the dates at which uncertainties are resolved may enter very strongly into an individual's preferences for alternative time streams of consequences. If the uncertainty is about whether one has a terminal illness, for example, some individuals might prefer to wait as long as possible before resolving it, while others would prefer to have it resolved right away. In other words, one's personal uncertainty or information state can be an important component of one's personal intertemporal utility function for alternative flows of consequences (Kreps and Porteus 1978).

TIME PREFERENCE AND RISK ATTITUDE: Where health and safety are at stake, individual attitudes toward time and risk may be confounded (Gafni and Torrance 1984). Moreover, both risk attitude and time preference can vary depending on the health or safety consequence at stake. For example, many people might prefer a monotonically decreasing hazard rate to a monotonically increasing one for injuries or diseases from which they will recover, on the grounds that they would prefer to get the high-risk regime over with as soon as possible. (In part, this may reflect a preference for early resolution of uncertainties.) On the other hand, the same people may prefer increasing to decreasing hazard rates for death or serious and irreversible illness or injury, on the grounds that it is desirable to postpone the high-risk regime for the onset of such irreversible changes as long as possible. Thus, individual time preferences and risk attitudes for risks to human health and safety depend on the effects at stake, and especially on whether they are reversible.

Construction of individual intertemporal utility functions that deal satisfactorily with these issues will require both conceptual and empirical advances. On the theoretical side, there have been some promising starts. For example, time streams of risks or benefits may in some circumstances be represented numerically by considering the constant (time-invariant) level of risk or consumption level, respectively, that is considered by the respondent to be "equivalent" to, or indifferent to, the time stream being evaluated. (See Barrager 1980, Shepard and Zeckhauser 1984, and Chew 1983 for examples of this strategy.) Also, generalizations of subjective expected utility theory to account for psychological aspects of the ways in which people think about risky prospects – and particularly about small-probability, large-consequence events – are being developed. (See, e.g., Quiggin 1982 and references therein.) In practice, however, individuals routinely violate normative conditions such as consistency, typically by undervaluing future consequences and hence regretting past choices even when past beliefs are vindicated. In addition, the loss in utility from irreversible changes (such as loss of limbs) tends to be underestimated prior to the fact, leading to risk-averting behavior that is seen in retrospect as having been inadequately cautious (Bodily 1980). Thus, "idealized" time preferences could be especially useful as a basis for guiding policy decisions that might be misled by holistic preference data. At the same time, development of a framework for defining and eliciting such idealized time preferences remains a matter for future research.

In principle, the difficulties and challenges in constructing satisfactory intertemporal utility functions for individuals can be bypassed, as a last resort, by using holistic evaluations (e.g., lottery-based assessments) of carefully described consequence time streams. As a practical matter, however, the complexity of time streams as cues for eliciting preference data makes such holistic judgments suspect and makes a systematic approach for decomposing the assessment of time streams particularly valuable.

Aggregation of Individual Utility and Value Functions for Time Streams of Consequences

The discount rate and aggregation of value functions

Once individual utility functions for uncertain flows of consequences have been estimated, either holistically or through a multiattribute/multiperiod decomposition and aggregation, the problem of aggregating them into an intertemporal social utility function arises. If consequences were deterministic, then the rationales given above for additive aggregation of individual utility functions would still apply, and the social utility for a known flow of consequences would be found by (a) having each individual assess his own value for the time stream, and (b) aggregating values across individuals via weighted summation. Note that even if each individual's intertemporal value function is derived as a discounted sum of single-period value functions, as in the Koopmans framework discussed above, aggregate social value is not achieved by applying an aggregate discount rate to the time stream of consequences from a regulation. Rather, it is obtained by aggregating the values obtained when each individual applies his own discount rate to the time stream.

From this perspective, seeking an aggregate discount rate to represent the social time preference rate represents the wrong level of aggregation. Preferences should be aggregated across time "within" individuals, and the results then aggregated "across" individuals. It is therefore not surprising that paradoxes emerge when a social discount rate is sought. For example, it may be that everyone would prefer alternative A to alternative B, where each alternative produces a flow of consequences, since differences in individual discount rates would allow lending and interest payment transactions to be worked out under A that would make everyone better off than under B. And yet there might be no single (aggregate or social) discount rate at which A would be preferred to B. Thus, using *any* aggregate discount rate would mislead social choice between the two alternatives (Sugden and Williams 1979). Of course, this argument only applies in contexts – including benefits assessment for health, safety, and environmental regulations – where there is no efficient market that allows decentralized adjustment of personal consumption and investment decisions until all personal time tradeoff rates are equated to a common market rate.

Choosing equity weights

If individual values are aggregated through weighted summation, then the equity weights for different individuals must be selected. As in the single-period case, considerations of relative need, relative contribution or sacrifice, and differences in individual risk attitudes might seem to be important; in addition, weights might be adjusted to reflect the expected remaining life spans of the stakeholders involved. How exactly these factors should enter into the determination of equity weights is not clear, however. As a practical matter, weights that are equal or that are proportional to expected remaining life years suggest themselves; as usual, the selected weights should be accompanied by a sensitivity analysis. Bodily (1980) considers the aggregation problem for health and safety programs.

Aggregation of Individual Intertemporal Utility Functions and the Need for Alternative Approaches

Risk sharing and utility aggregation

In principle, individual utilities for time streams of consequences may be aggregated like any other utilities – for example, through weighted summation. But this simple statement hides a variety of practical problems, having to do with the fact that subsequent individual decisions help determine what the consequences of a regulatory decision will be and how the consequences will be distributed. Thus, individual utilities for consequence streams no longer translate directly into utilities for regulatory acts.

Aggregating individual preferences is especially difficult when the time stream of consequences from a regulation is uncertain. Indeed, just as there may be no single aggregate discount rate (in the absence of perfect capital markets) that represents group preferences for acts if the group's members can use mutually agreeable loans to exploit differences in their individual discount rates, so no social utility function for acts (e.g., regulatory alternatives) may exist that can represent group preferences for acts if the group's members can use voluntary risk-sharing transactions to exploit differences in their utility functions. (Here, group preference is assumed to coincide with individual preferences when all individuals agree; see Raiffa 1968: Chapter 8 for details.)

The question of whether a particular risk/benefit combination should be accepted by a group cannot be separated from the question of how risks and benefits should be shared among the group's members – and, in particular, of whether there is any way of partitioning the risk/benefit combination into shares so that each member of the group finds his share acceptable (in the sense that he would prefer having it to not having it; see e.g., Raiffa 1968; Eliashberg and Winkler 1981). Indeed, important qualitative properties of group utility functions, such as whether society should be more or less risk-averse in its investment decisions than are its members, will depend on the possibilities for risk sharing, as well as on the utility functions of the individual members. In general, effective risk sharing will make a group less risk-averse than its members in the risks that it is willing to accept (Arrow and Lind 1970). On the other hand, if outcomes to all members are strongly and positively correlated (as might be the case for a safety measure reducing the probability of a catastrophic accident), then the group may be more risk-averse than its members.

The pattern of correlations across outcomes to different individuals forms the basis for distinguishing between *routine* (statistically independent, usually single-fatality) risks, such as those from automobile accidents or many occupational hazards, and *catastrophic* (large, multi-fatality) risks, such as those from industrial facility explosions or large-scale releases of hazardous material, for example, from derailment of a train carrying chlorine. It is well known that most individuals value reductions in catastrophic risk more than reductions in routine risks that are expected to save a comparable number of lives, perhaps because catastrophic risks carry a connotation of being imposed on a large number of people simultaneously, while routine risks are more apt to reflect voluntary individual choices regarding what jobs to take, how fast to drive, and so forth (see, e.g., Starr and Whipple 1984). Thus, an important attribute to be included in the description of regulatory benefits, in eliciting individual preference data, is the pattern of correlations among outcomes to different individuals.

In summary, the possibility of risk sharing implies that group preferences for alternative risky prospects – and hence the social benefit or utility assigned to alternative acts yielding uncertain consequences – cannot in general be assessed without considering the distribution of shares in the outcome among individuals, and hence the risk-sharing process. Put otherwise, the social utility from an act can only be determined from consequences for individuals, which in turn will depend on how those individuals have agreed to partition the uncertain consequences through insurance arrangements, contingent contracts, and so forth. The correlation pattern among outcomes to different individuals may also itself be an important argument in individual utility functions, reflecting a concern for equity, for the societal impact of a regulation, or for the welfare of others (Raiffa, Schwartz, and Weinstein 1977).

Such risk-sharing transactions are more likely to be possible in the context of pure financial risks than in the context of health, safety, and environmental regulations. Money is transferable; health is not. The problem of assessing benefits is not so difficult when individuals have no control over the distribution of impacts. But even the benefits from a health and safety regulation may depend on individual insurance policies, workman's compensation plans, and other nonregulatory measures taken to cope with risks prior to the introduction of the regulation.

The nonexistence of social utility functions for acts presents no stumbling block for the development of social utility functions for consequence streams. However, consequence assessment does abstract away from the (possibly complicated and politically important) processes, such as risk-sharing and compensation schemes, through which consequences for individuals are generated. And these processes must be considered anyway in order to forecast the probable consequences to individuals of alternative policies. Moreover, if active *participation* in political decision, risk-sharing, and collective choice processes is itself a source of benefits to individuals – a position advanced by several early political economists – then abstracting away from such processes to the consequences (in terms of time streams of health, safety, environmental, and economic attributes) that they produce provides an inadequate basis for regulatory benefits assessment. The same distributions of tangible consequences may be assessed quite differently, depending on how they were arrived at, and specifically on details such as the extent of prior informed consent, active individual participation in the decision process, and agreement to accept the outcome.

The timing of benefits assessment: *ex ante* versus *ex post* benefits

It is also necessary to distinguish between group utilities for acts *prior* to the resolution of uncertainties and group utilities *following* the resolution of uncertainties. For example, two individuals with divergent beliefs and risk attitudes may both gain in terms of their prior expected utilities by betting, if each expects to win the bet. Yet only one can actually win, so both cannot gain *ex post*. This sort of situation can be important in the analysis of health and safety regulations, for example when some individuals are in effect willing to "bet" that there is no need for a safety regulation (e.g., a ban on a suspected carcinogen or on the siting of a hazardous facility) and hence would prefer not to have it. Insurance or compensation schemes, especially those that involve payment of a prior risk premium in exchange for an assumption of risk by the paid party, are in effect just such bets. (Note, however, that both insurer and insured can gain in the aggregate, *ex ante* and *ex post,* when many independent risks are insured, so that there is risk sharing as well as pure betting involved.) In assessing the benefits of a risk-reducing regulation, it is important to specify whether the assessment is made *ex ante* or *ex post* – or, if uncertainties are only gradually resolved, as is usually the case, what information is to be used or assumed in making the assessment.

A closely related point has to do with the assessment of benefits from a regulation that prevents an accident (or accidents) that might not have occurred anyway even in the absence of regulatory intervention. Safety regulations that reduce the *probability* of an accident have this characteristic. The value of such a regulation is seldom known *a priori,* and may not even be known *ex post,* if it is never revealed whether an accident would have occurred had it not been for the regulation. Such cases strain the bounds of the current state of the art of benefits assessment. Although in principle prior subjective expected utilities can be elicited for risk reductions and then aggregated as for any other benefit, this does not provide a very satisfactory practical approach when respondents are unfamiliar and inexperienced with the risks and consequences involved, poorly calibrated in their understandings of probabilities, or when state-dependent utilities are important. In any of these cases, utilities assessed *ex post* may differ substantially from prior utility estimates. Moreover, market-based approaches – which rely on reductions in insurance premiums and other risk-avoidance costs due to the regulation – provide very imperfect guidance where life and health status, for which *ex post* monetary equivalents may be undefined, are at stake.

In summary, the time frame or information state with respect to which welfare analyses are to be conducted must be clearly specified, and the specifiction adopted may influence the outcome of the analysis.

The collective nature of dynamic decisions

As previously mentioned, the value of a decision made now is often contingent on what subsequent decisions are made. More generally, the value of a current decision, such as a regulation requiring that seatbelts be worn, may depend on how other decision makers, present or future, respond to it.

The fact that the consequences of current acts – and hence individual preferences among acts – may depend on the choices of others raises some special problems for aggregation of individual preferences for acts. For example, each individual might in principle be able to identify a most-preferred policy (sequence of acts conditioned on observations), and all these policies might require the same first act. Thus, each individual would choose the same first act if he could dictate society's choice of policy. Yet, if this act was also the first step in a very undesirable policy, and if individuals were uncertain about what future acts might be chosen, then they might prefer some other act to be implemented. In effect, current decision makers may end up trying to second-guess or play strategic games with future decision makers.

Two aspects of this problem have already been mentioned in Chapter 4 for decisions over time. One is that individuals may unanimously wish society to use a lower discount rate in assessing future benefits than any single individual would use. For example, each individual may feel willing to save for future generations at a greater rate if he is assured that everyone else will do so too; this can be accomplished, for example, by using a low collective discount rate in centralized planning for the extraction of nonrenewable resources. Secondly, each generation may end up saving less for distant future generations than it would if it could be assured that intermediate generations would cooperate by not reaping all of the potential increases in consumption rates for themselves (Sen 1967).

Theoretical niceties aside, it is extremely important to note that individuals may unanimously prefer society in its collective decisions to be longer sighted than any individual would be in his own decisions. In eliciting preference data for guiding public decisions it is therefore imperative to distinguish clearly between an individual's preferences for collective actions in which he participates and his preferences for his own actions in isolation.

Summary of Social Utility Perspective

The preceding discussions have described in some detail the social utility function approach to benefits assessment. Figure 5–2 provides an overview of various components of the social utility framework. The basic logic of the approach is simple. Individual changes in welfare from proposed alterna-

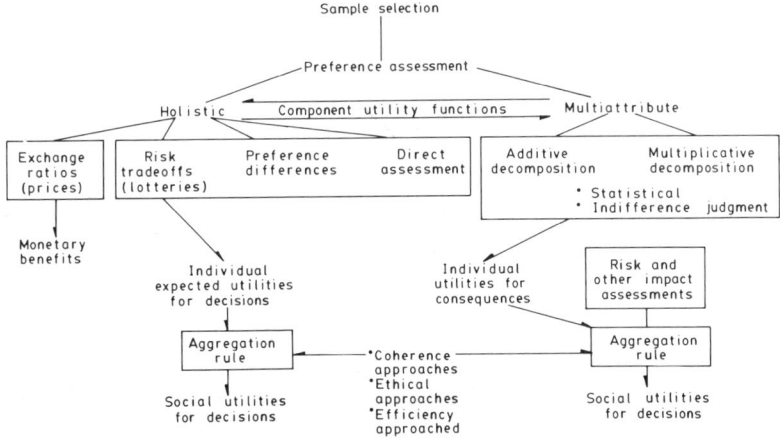

Fig. 5–2. Overview of the social utility framework.

tives are estimated using individual utility functions over relevant attributes, or holistic utility or value judgments, if these can be made reliably. The results are expressed in terms of utilities, rather than in monetary terms, so that they are not confounded by budget constraints on monetary income and do not require the usual assumptions of divisibility, excludability, and transferability that are needed for conventional interpretations of market prices. (This does not mean that individual choice behavior is unconstrained; rather, the constraint is that one must choose directly among alternatives and among lotteries over alternatives.) It is assumed that the value or utility scale on which individual benefits are measured is a difference scale (i.e., values are measurable, and strength of preference has meaning). Finally, the changes in individual utilities are aggregated, usually through unweighted addition, after they have been scaled to lie on a common axis, for example, through the use of common endpoints and an ethical (and empirically untestable) assumption that the utility difference between endpoints is the same for all individuals – or at least should be treated as if it were the same.

These developments can now be used to answer many of the questions raised at the beginning of this chapter from the standpoint of the social utility framework. Specifically, the following general guidelines can be offered.

Q1. How shall the benefits to each stakeholder group be defined – subjectively, in terms of their preferences, or objectively, in terms of measurable changes in physical and economic variables?

A1. Benefits should be *described* in terms of objective variables, namely, the set of attributes or probability distributions over attributes. If preferences are known to be monotonically increasing or decreasing in attribute levels, then this objective level of description may be sufficient to allow some inferior alternatives to be eliminated via dominance or stochastic dominance (see, e.g., Zeleny 1982: Chapter 11 and references therein; also Roy 1977 for various extensions of the dominance concept). To obtain quantitative *evaluations* of individual benefits, however, individual preferences, as summarized in individual value or utility functions, should be used. Thus, individual welfares should be expressed in terms of values or utilities; that is, by comparison to a most desirable outcome and a least desirable outcome. For example, the value assigned to a particular outcome by a homogeneous stakeholder group may be thought of as the percent of the potential value gain between that group's least- and most-preferred outcomes that is obtained by the actual outcome. The units in which benefits are finally expressed are thus units of relative desirability, rather than more concrete units such as dollar equivalents, increase in life expectancy, and so forth. This response should be contrasted to the answer given in the econometric approach, which is that benefits should be measured in terms of individual willingness to pay.

Q2. How should benefits to a group be aggregated across time and in the

5. Theory of Regulatory Benefits Assessment

presence of uncertainties? What attitude toward risk should be taken in the aggregation, and how should it depend on the risk attitudes of the stakeholders involved?

A2. From the social utility perspective, each individual should assess his own value or utility for time streams of consequences. The resulting values or utilities should then be aggregated across individuals in the usual way (see Q3 below). If outcomes are correlated across individuals, then the pattern of correlation should be included in the descriptions of alternatives provided to individuals in eliciting their preferences. Attitudes toward risk are thus incorporated at the level of individual utility functions, rather than having to be decided on after the decentralized preference data have been collected. In general, positively correlated individual risks will lead to social risk aversion, whereas uncorrelated (statistically independent) individual risks will lead to social risk neutrality, at least if effective risk sharing is possible. The institutional possibilities for risk sharing may strongly affect group attitudes toward risk, as well as the distribution of consequences to individuals.

The problem of how to incorporate the possible preferences of future generations remains a matter for policy decision and may require elicitation of individual preferences for collective (social) policies regarding preservation of resources and attitudes toward activities with irreversible consequences. In other words, as with the assessment of risks, the policy question of what attitude to take toward future generations may be pushed down to the level of individual preferences—but individual preferences about collective decisions.

No adequate theory of intertemporal utility assessment for risks involving irreversible catastrophic changes (e.g., death or crippling injury) currently exists, even at the level of individuals, although there is a considerable literature on utility functions for consumption/investment time streams. At the moment, risky consequence streams involving potential health and safety damages must in general be assessed holistically.

Q3. How should benefits be aggregated across stakeholder groups, especially when some groups gain and others lose?

A3. The answer given by social utility theory is that under a fairly wide range of normative conditions aggregation should proceed through (possibly weighted) summation. However, this approach presupposes the legitimacy of interest of all the parties whose preferences are represented in the sum. For example, there is no protection in this rule against tyranny of the majority, except that the distribution of outcomes across individuals may be one argument in the utility functions of individuals, thus allowing their ethical and distributional concerns to be expressed. Unless information about correlations among individual outcomes and similar distributional information is provided and accounted for in the elicitation of individual preferences among alternatives, the weighted sum aggregation rule may prove ethically unacceptable. Once again, this is consistent with the philosophy of pushing policy decisions down to the level of individual preferences regarding collective matters, in denial of the principle that efficiency and equity considerations can usefully be separated when collective decisions, rather than market goods, are at stake.

If weighted summation is used to aggregate individual values or utilities, then the weights should be varied, to provide a sensitivity analysis of the results. In general, varying the weights (assumed to be non-negative and normalized to sum to one) singles out different Pareto-efficient outcomes.

It should be noted that there are a number of alternatives to aggregation, involving negotiated compensation agreements, insurance arrangements, and so forth, for reconciling differences in individual utilities and reaching a collective decision. The essence of the benefits-assessment procedure. however, is that it seeks to measure benefits, rather than concerning itself with the process through which benefits are obtained.

Q4. How are preferences of different qualities (i.e., based on different amounts of reflection and information) to be treated?

A4. Although the social utility approach allows holistic assessment of individual preferences, the theory of multiattribute utility assessment allows preferences for consequences in terms of attributes to be assessed separately from beliefs about what changes in attribute levels will result from any particular proposed alternative. Thus individuals who are poorly informed about the probable consequences of regulatory alternatives can still have their preferences for consequences and for lotteries over consequences represented. It then falls to the analyst and public decision maker to assess the probable consequences of each alternative as well as possible. Alternatively, if there is no special expertise that can be brought to bear on the problem, then holistic assessments of the utilities for alternative actions (rather than for consequences) can be assessed and aggregated; these involve beliefs as well as preferences. If individual information is very diverse, however, then it may be possible to achieve *ex ante* gains through contingent contracts, for example, by making the eventual financing of a regulation (the distribution of compliance costs between producers and consumers in a regulated market) contingent on the outcome obtained.

Advantages of the social utility approach

Despite the limitations of the social utility approach in ignoring the processes through which benefits can be obtained, it does offer potentially useful solutions to some of the weaknesses of traditional benefit-cost analyses outlined in Chapter 2. Specifically, it offers a way of quantifying benefits that are traditionally hard to quantify – for example, those involving pain and suffering – namely, through use of values or utilities, which allow multiattributed alternatives to be assessed on a single commensurable scale. Secondly, it allows concerns about equity, distribution of costs

5. Theory of Regulatory Benefits Assessment

and benefits, and correlations in the outcomes to different individuals under risk to be taken into account (as arguments in individual utility functions), since more than just the flow of economic goods and consequences across the household boundary can be considered when preference data for collective decisions or consequences are elicited. In addition, it offers a way of quantifying individual benefits that is less sensitive to artifacts of income distribution than most traditional approaches, and that can be used even where inadequate market data are available, since surveys, interviews, or questionnaires are the primary sources of data. Indeed, the social utility approach avoids the problems with partial-equilibrium methods of analysis, such as the difficulties in assessing nonmarginal changes. This is not because it is a general-equilibrium approach, but because it is a *nonequilibrium* approach. The rationales for assessment and aggregation in the social utility framework do not require or suppose the invisible hands of market mechanisms but rather provide an alternative to the market mechanism for valuing goods – including collective or nonmarket goods – directly. The social utility framework would thus appear to be most useful in situations where market-based approaches are weakest, namely, in the assessment of benefits from changes in the levels of nonmarket goods.

Table 5–3 summarizes several of the main features of both the econometric and the social utility frameworks. Although the table has been arranged to facilitate comparison of these two stereotyped approaches, it should be recognized that in many applications they are complementary, and aspects of each may usefully be brought to bear in a single benefits assessment.

Table 5–3. Summary of econometric and social utility frameworks

	Econometric framework	Social utility framework
Assumptions	Dollar measures of value	Utilities, value scale measure of value
	Market equilibrium rationale/interpretation	Psychometric rationale/interpretation
	Preferences transferable across contexts	Axiomatic basis/interpretation for aggregation
	Separation of equity and efficiency	Decentralized equity judgments
	Interpersonal monetary comparisons of value	Interpersonal utility comparisons of value
	Revealed preferences	Announced preferences
	Informed choice	Coherent choice

Table 5-3 (continued)

	Econometric framework	Social Utility framework
Data Requirements	Market price data Sufficiently rich stimulus and response sets	Expressed preference data Binary Ordinal Cardinal
	Uses market data	Uses direct preference data
Strengths	Measurements are objective	Interaction effects, nonmarginal changes
	Avoids strategic misrepresentation	Avoids strategic misrepresentation
	Flexibility and validation of price imputation	Expertise separated from preferences
	Direct implications for financing	Coverage of relevant attributes Treatment of uncertainty
Weaknesses	Preferences for collective choices, collective consequences, nonuse benefits, not well expressed	Hypothetical bias
	Income confounding	
Appropriate Domains of Application	Air quality	Air quality
	Occupational risks	Public health and catastrophic risks
	Assumed (first-party) risks	Imposed (third-party) risks
	Static choice	Static choice

Arthur D. Little, Inc.

References

Abonyi, G. 1983. 'Filtering: An Approach to Generating the Information Base For Collective Action.' *Management Science* 29 (4).

Amershi, A. and J. H. W. Stoeckenius. 1983. 'The Theory of Syndicates and Linear Sharing Rules.' *Econometrica* 51 (5): 1407–1416.

Arrow, K. J. 1963. *Social Choice and Individual Values*. 2nd ed. New Haven: Yale University Press.

——— and R. C. Lind. 1970. 'Uncertainty and the Evaluation of Public Investment Decisions.' *American Economic Review* 60: 364–378.

Banzhaf. J. F. III. 1968. 'One Man, 3.312 Votes: A Mathematical Analysis of the Electoral College.' *Villanova Law Review* 13: 304–332.

Barrager, S. M. 1980. 'Assessment of Simple Joint Time/Risk Preference Functions.' *Management Science* 26 (6): 620–632.

Bean, J. C. and R. L. Smith. 1984. 'Conditions for the Existence of Planning Horizons.' *Mathematics of Operations Research* 9 (3): 391–400.

Bell, D. E. 1977. 'A Utility Function for Time Streams Having Inter-period Dependencies.' *Operations Research* 25 (3): 448–458.

———. 1983. 'Risk Premiums for Decision Regret.' *Management Science* 29 (10).

———, R. L. Keeney, and H. Raiffa, eds. 1977. *Conflicting Objectives in Decisions.* New York: Wiley.

Bergstrom, T. C. and R. C. Cornes. 1983. 'Independence of Allocative Efficiency from Distribution in the Theory of Public Goods.' *Econometrica* 51 (6): 1753–1765.

Bertsekas, D. P. 1976. *Dynamic Programming and Stochastic Control.* New York: Academic Press.

Bodily, S. E. 1979. 'A Delegation Process for Combining Individual Utility Functions.' *Management Science* 25 (10).

———. 1980. 'Analysis of Risks to Life and Limb.' *Operations Research* 28 (1): 156–175.

Brock, H. W. 1980. 'The Problem of Utility Weights in Group Preference Aggregation.' *Operations Research* 28 (1): 176–187.

Campbell, D. T. and J. C. Stanley. 1963. *Experimental and Quasi-Experimental Designs for Research.* Chicago: Rand McNally.

Chew, S. H. 1983. 'A Generalization of the Quasilinear Mean with Applications to the Measurement of Income Inequality and Decision Theory Resolving the Allais Paradox.' *Econometrica* 51 (4): 1065–1092.

Cochran, W. G. 1953. *Sampling Techniques.* New York: Wiley.

Currim, I. S. and R. K. Sarin. 1984. 'A Comparative Evaluation of Multiattribute Consumer Preference Models, *Management Science* 30 (5): 543-561.

Dyer, J. S. and R. K. Sarin. 1979. 'Group Preference Aggregation Rules Based on Strength of Preference.' *Managment Science* 25 (9).

Eliashberg, J. and R. L. Winkler. 1981. 'Risk Sharing and Group Decision Making. *Management Science* 27 (11).

Epstein, L. G. 1983. 'Stationary Cardinal Utility and Optimal Growth Under Uncertainty.' *Journal of Economic Theory* 31 (1)

Fishburn, P. C., 1967. 'Methods of Estimating Additive Utilities.' *Management Science* 13: 434–453.

———. 1984. 'Equity Axioms for Public Risks.' *Operations Research* 32 (4): 901–908.

Freeman, M. A. 1979. *The Benefits of Environmental Improvement.* Baltimore: Johns Hopkins University Press.

Gafni, A. and G. W. Torrance. 1984. 'Risk Attitude and Time Preference in Health.' *Management Science* 30 (4): 440–451.

Green, J. R. and J. J. Laffont. 1979. *Incentives in Public Decision Making.* New York: North-Holland.

Harsanyi, J. C. 1952. 'Cardinal Welfare, Individualistic Ethics, and Interpersonal Comparisons of Utility.' *Journal of Political Economy.* Reprinted in E. S. Phelps, ed., *Economic Justice.* Baltimore: Penguin, 1973.

Hershey, J. C., H. Kunreuther, and P. J. H. Schoemaker. 'Sources of Bias in Assessment Procedures for Utility Functions.' *Management Science* 28 (8): 936–954.

Hildenbrand, W. 1983. 'On the Law of Demand..' *Econometrica* 51 (4): 997–1019.

Holmstrom, M. and R. B. Myerson, 1983. 'Efficient and Durable Decision Rules with Incomplete Information.' *Econometrica* 51 (6): 1799–1819.

Howard, R. 1984. 'On Fates Comparable to Death.' *Management Science* 30 (4): 407–422.

Horsky, D. and M. R. Rao. 1984. 'Estimation of Attribute Weights from Preference Comparisons.' *Management Science* 30 (7): 801–822.

Kahneman, D., P. Slovic, and A. Tversky, eds. 1982. *Judgement Under Uncertainty: Heuristics and Biases*. New York: Cambridge University Press.

Karni, E., D. Schmeidler, and K. Vind. 1983. 'On State Dependent Preferences and Subjective Probabilities.' *Econometrica* 51 (4): 1021–1032.

Keeney, R. and C. Kirkwood. 1975. 'Group Decision Making Using Cardinal Social Welfare Functions.' *Management Science* 22 (4)' 430–437.

———, and H. Raiffa. 1976. *Decisions with Multiple Objectives: Preferences and Value Tradeoffs*. New York: Wiley.

Kirkwood, C. 1979. 'Pareto Optimality and Equity in Social Decision Analysis.' *IEEE Transactions on Systems, Man, and Cybernetics*. SMC-9 (2): 89–91.

Knetsch, J. L. and J. A. Sinden. 1984. 'Willingness to Pay and Compensation Demanded: Experimental Evidence of an Unexpected Disparity in Measures of Value.' *Quarterly Journal of Economics* (August).

Krantz, D. H., R. D. Luce, P. Suppes, A. Tversky. 1971. *Foundations of Measurement* Vol. I. New York: Academic Press.

Kreps, D. M. and E. L. Porteus. 1978. 'Temporal Resolution of Uncertainty and Dynamic Choice Theory.' *Econometrica* 46 (1): 185–200.

Krischer, J. P. 1980. "An Annotated Bibliography of Decision Analytic Applications to Health Care." *Operations Research* 28 (1).

Kunreuther, H., J. Linnerooth, and J. W. Vaupel. 1984. 'A Decision-Process Perspective on Risk and Policy Analysis.' *Management Science* 30 (4): 475–485.

———, and T. E. Morton. 1973. 'Planning Horizons for Production Smoothing with Deterministic Demands.' *Management Science* 20 (1): 110–121.

Luce, D. R. and H. Raiffa. 1957. *Games and Decisions*. New York: Wiley.

March, J. G. 1966. 'Party Legislative Representation as a Function of Election Results.' In P. F. Lazarsfeld and N. W. Henry, eds., *Readings in Mathematical Social Science*. Cambridge, Mass.: M.I.T. Press.

Merkhofer, M., V. Covello, J. Menkes, eds. 1984. *Risk Assessment and Risk Assessment Methods*. Report prepared for the National Science Foundation. Boston: Charles River Associates.

Miller, J. C. and B. Yandle, eds. 1979. *Benefit-Cost Analysis of Social Regulation*. Washington, D.C.: American Enterprise Institute for Public Policy Research.

Montgomery. 1976. *Design and Analysis of Experiments*. New York: Wiley.

Pliskin, J. S. and C. H. Beck, Jr. 1976. 'A Health Index for Patient Screening: A Value Function Approach with Application to Chronic Renal Failure Patients.' *Management Science* 22 (9): 1009–1021.

———, Shepard, D. S., and M. C. Weinstein. 1980. 'Utility Functions for Life-years and Health Status.' *Operations Research* 28 (1): 206–224.

Quiggin, J. 1982. 'A Theory of Anticipated Utility.' *Journal of Economic Behavior and Organization* 3 (4).

Raiffa, H. 1968. *Decision Analysis*. Reading: Addison-Wesley.

———, W. B. Schwartz, and M. C. Weinstein. 1977. 'Evaluating Health Effects of Social Decisions and Programs.' In *Decision Making in the Environmental Protection Agency*. Washington, D.C.: National Academy of Sciences.

Roberts, F. S. 1979. *Measurement Theory, with Applications to Decisionmaking, Utility, and the Social Sciences*. Reading: Addison-Wesley.

Roy, B. 1977. 'Partial Preference Analysis and Decision Aid.' Chapter 2 in Bell *et al.* (1977).

Sen, A. K. 1967. 'Isolation, Assurance, and the Social Rate of Discount.' *Quarterly Journal of Economics* 81.

———. 1970. *Collective Choice and Individual Welfare*. Edinburgh: Oliver and Boyd.

Shepard, D. S. and R. Zeckhauser. 1984. 'Survival vs. Consumption.' *Management Science* 30 (4): 423–439.
Starr, C. and C. Whipple. 1984. 'A Perspective on Health and Safety Risk Analysis.' *Management Science* 30 (4): 452–463.
Stokey, E. and R. Zeckhauser. 1978. *A Primer for Policy Analysis*. New York: Norton.
Sugden, R. and A. Williams. 1979. *The Principles of Practical Cost-Benefit Analysis*. Oxford: Oxford University Press.
Torrance, G. W., M. H. Boyle, and S. P. Horwood. 1982. 'Application of Multiattribute Utility Theory to Measure Social Preferences for Health States.' *Operations Research* 30 (6): 1043–1069.
Tversky, A. 1979. 'On the Elicitation of Preferences: Descriptive and Prescriptive Considerations.' Chapter 9 in Bell *et al.* (1977).
Von Winterfeldt, D., and W. Edwards. In press. *Decision Analysis and Behavioral Research*.
Weinstein, M. C., D. S. Shepard and J. S. Pliskin. 1980. 'The Economic Value of Changing Mortality Probabilities: A Decision-Theoretic Approach.' *Quarterly Journal of Economics* (March): 373–396.
Whittle, P. 1982. *Optimization over Time: Dynamic Programming and Stochastic Control*. Vol. I. New York: Wiley.
Zeleney, M. 1982. *Multiple Criteria Decision Making*. New York: McGraw-Hill.

Chapter 6

General Methods for Benefits Assessment

Ronald G. Cummings, Louis Anthony Cox, Jr., and A. Myrick Freeman III

The preceding chapters have discussed a number of general approaches and methodological frameworks for assessment of health, safety, and environmental benefits, and have examined their conceptual and philosophical foundations in some detail. In this chapter we close our discussion of generic methods with an overview of some specific approaches and techniques that have proven useful in recent benefits assessments.

Because the conceptual and philosophical groundwork for these methods has already been laid, our primary purpose here is to describe a number of practical techniques of benefits assessment that are now being used, or that have been proposed for use, and to describe their principal strengths, weaknesses, assumptions, and appropriate areas of application. That is, we attempt to provide a useful catalogue of methods to assist the practitioner in selecting the basic approach that seem most appropriate for his needs in carrying out specific benefits assessments.

Our emphasis throughout is on *monetary evaluation of nonmarket goods* (e.g., public and open-access goods). This choice of scope reflects the facts that

1. Excellent summaries of methods for assessing the benefits associated with *market* goods, such as changes in prices (including costs, factor rents, and wages) or changes in quantities produced, are already available (see, e.g. Freeman 1979b: Chapters 3 and 4).

2. Nonmonetary expressions of benefits have already been discussed in Chapter 5. Although many of the benefits assessment techniques surveyed in this chapter are compatible with benefits expressed in terms of aggregate utility rather than dollars, dollar measures have conventionally been used for such purposes. We therefore have little to add to the dicussion in Chapter 5 in terms of specific suggestions for implementation of nonmonetary methods.

Finally, in this chapter we concentrate on a prescriptive outline of methods and approaches to implementation and do not try to describe or critically appraise current agency practices associated with these methods. For critical descriptions of several case studies involving regulatory benefits assessments, the reader may consult Miller and Yandle (1979).

As mentioned in the preceding chapters, techniques for obtaining monetary values of nonmarket goods can be broadly categorized as relying either on observed behavior and choices or on responses to hypothetical situations posed to individuals by interviewers or printed questionnaires. The first category includes all techniques relying on observed demand functions or cost functions, changes in prices of goods or factor inputs, or observed changes in some nonmarket activity such as recreation (Freeman 1979a). Examples include the use of property value differentials, household expenditures (on cleaning or on maintenance and repair of materials damaged by pollutants), and travel costs incurred to participate in recreation. These approaches are often referred to as *indirect* methods because they involve efforts to infer monetary values by using the relevant economic models to interpret data on individual behavior.

The second category of benefit estimation techniques includes asking people directly about values, as in willingness-to-pay surveys, bidding games, and contingent valuation surveys. This category also includes those techniques in which people are asked how their behavior would change with a change in environmental quality (e.g., would you choose a different recreation site?) or are asked to rank alternative scenarios involving different bundles of environmental quality and other attributes (contingent ranking studies). These techniques all have in common the fact that the choice or value questions involve hypothetical situations. Thus, the reliability of such methods is contingent upon a close correspondence between how people would actually choose in a particular situation and how they say they would choose when asked by the interviewer.

The following sections are devoted to a review of several direct and indirect techniques for estimating market and individual use benefits. The concepts of nonuse benefits and some problems in their estimation are discussed in Chapter 8. Given our goal of providing a useful catalogue of techniques and enough information to help the practitioner choose among them, our treatment of each is necessarily brief. However, at the end of the chapter there is an extensive bibliography, which includes references noted in the text on the conceptual background, details of implementation, and practical experience to date for each method.

Despite the summary nature of the chapter, we have devoted a disproportionate amount of attention to one approach – the hedonic price method – that seemed to us to be exceptionally important. This is because this method best represents the generic econometric approach described in Chapter 5, which we see as being the dominant paradigm for non-survey-based benefits assessments.

6. General Methods for Benefits Assessment

Direct Monetary Methods for Evaluating Public Goods

The Contingent Valuation Method

The contingent valuation method (CVM) involves the use of surveys as a means for deriving estimates for social benefits attributable to public or other nonmarket (e.g., open-access) goods. The CVM, as it is typically applied, draws upon a market analogy in that the survey is designed around the notion of a contingent market (Randall *et al.* 1978; Randall, Hoehn, and Tolley, in press). Thus the survey participant has described to him a proposed change in the level at which a public good is provided; for example (i) a reduction in air pollution or (ii) an increase in air pollution. Then the subject is asked the maximum amount he would be willing to pay for the proposed change in the case of (i), or the minimum amount that he must be paid to accept it or the maximum amount he would pay to prevent it in the case of (ii). Procedural steps for applying the CVM are as follows:

1. Design of the survey instrument (questionnaire), which includes the following key features (in general, see Dasgupta and Pearce 1978; Cummings *et al.* 1984):
   a. Cogent explanation of the survey and its purposes.
   b. Description of the public good which is to be valued.
   c. Request for subjects' income and expenditure patterns (optional).
   d. The willingness-to-pay (or willingness-to-accept compensation) question, which includes description of payment vehicle.
   e. Interactive value questions (optional) given subject's response to d, questions of the sort: If, with all households paying (*response to d*), the public good cannot be provided, would you pay (accept) $1.00 more (less) to obtain it?
   f. Request for demographic (e.g., income, age, sex) and attitudinal data (optional).

2. Pre-tests of the survey instrument; when required, modifications of the instruments.
   a. Pilot study using a small selected sample of respondents and/or focus groups.

3. Choice of final sampling design and survey area (see Miller and Plott 1983).

4. Training of interviewers (unless mail survey is used).

5. Implementation of the survey (through interviewers or by mail).

Secondary data are not directly used in the CVM. For the sampling design (step 3 above) census data are often used for identifying households (or neighborhoods) to be included in the sample.

Output and interpretation

The output from applications of the CVM consists of willingness-to-pay (WTP) or willingness-to-accept (WTA) compensation values, and corresponding demographic information, for the N individuals included in the survey. If V_{ij} is the ith subject's valuation (WTP or WTA) of the jth increment in the provision of the public good in question, and D_{ik} is the value of the kth demographic/attitudinal variable for subject i, survey data obtained in the application of the CVM are typically analyzed by means of regression techniques exemplified in the simplest way by the following linear system:

$$V_{ij} = \sum_{i=1}^{n} \alpha_{jk} D_{ik} + \alpha_{ok}$$

Adjustments are typically made for "outliers" (Dasgupta and Pearce 1978; Desvousges and Smith 1982); logit techniques are often employed to adjust for nonrespondents, protest, and zero bids (Desvousges and Smith 1982; Bishop, Heberlein, and Kealy 1983; Bockstael and McConnell 1983; Miller and Plott 1983). For the set of variables D_k found to be statistically significant in explaining variations in V, measures $V_j(D_k)$ are desired which are interpreted as follows: for households with characteristics D_k, $V_j(D_k)$ is the household's maximum willingness to pay (accept) for the posited jth incremental change in the level at which the public good is provided, *as described* in the CVM survey. The aggregation across households with characteristics D_k of values $V_j(D_k)$ is used as a means for estimating social benefits (typically, consumer surplus) attributable to the described jth incremental change in the public good. (The aggregation is necessarily an inexact approximation since Hicks-compensated demand curves [Baulding 1975] are not available; at best, the method is used to simulate ordinary demand curves.) The array of values $V_1, V_2, \ldots$ approximate the Bradford bid curve (Bradford 1970) from which the values of marginal changes in the public good may be estimated.

Major assumptions

Two major behavioral assumptions underlie the use of the CVM:

a. Subjects can (and have incentives to) determined their preference orderings between the public and all other relevant goods and services (Feather 1959; Freeman 1979; Cummings *et al.* 1984).

b. Subjects will not behave strategically, that is, offer willingness-to-pay (or accept) values which, rather than reflecting their true preferences, are intended to bias survey results in directions which are consistent with the subject's preferences (Samuelson 1954; Bohm 1971, 1972; Smith 1976, 1977; Rowe and Chestnut 1983).

6. General Methods for Benefits Assessment

Strengths and weaknesses of the method

It must be recognized that the CVM is at a relatively infant stage of development and that considerable controversy exists as to the extent to which the method yields meaningful measures for public-good benefits. The major *weaknesses* of the method are as follows (for a more detailed overview, see Cummings *et al.* 1984);

1. The possibility of strategic bias (see assumption b above).

2. Biases and undesired influences upon responses, due to aspects of the survey design, such as starting bids, payment vehicles, information, question framing, and subject-interviewer interactions. These are especially apt to raise problems when respondents are uncertain about their "true" WTP values.

3. Given that WTA (WTP) amounts are hypothetical – they are not actually paid (received) – subjects may lack incentives to perform the introspective preference research process required to determine their true preference orderings (assumption a above). This problem may be exacerbated in instances where subjects are unfamiliar with the public good in question (Feather 1969; Slovic and Lichtenstein 1968; Bishop, Heberlein, and Kealy 1983).

4. Iterative bidding may bias results by "bullying" subjects into stating higher bids (see Mitchell and Carsons' comments in Cummings *et al.* 1984).

5. Questions exist as to the extent to which CVM values reflect attitudes as opposed to intended behavior (Ajzen and Fishbein 1977; Bishop, Heberlein, and Kealy 1983; Bockstael and McConnell 1983; Burness *et al.* 1983; Schulze *et al.* 1983).

The major *strengths* of the CVM follow:

1. The independence of the method from secondary data allows its (potential) application to a wide range of public and open-access goods (Chew and MacGrimmon 1979).

2. Existing evidence belies the existence of strategic bias in the CVM (weakness 1 noted above). Well-designed surveys may reduce the structural weakness maintained in 2 (Freeman 1979; Cummings *et al.* 1984).

3. While there is currently no accepted standard by which the accuracy of CVM values might be assessed, substantial evidence exists which demonstrates that under conditions described below under appropriate applications, the CVM generates values which usually compare well (are consonant with) analogous values obtained from alternative,

market-based methods (see Chapters 6 and 8 in Cummings *et al.* 1984). This "strength" is weakened, however, by the absence of knowledge as to the accuracy of values derived with these alternative methods.

4. Some evidence exists that suggests that interrogated, hypothetical values, such as those obtained in the CVM, outperform the standard expected utility model in predicting equilibrium prices determined in laboratory experiments, a finding which may mitigate the weight of weakness 3 (see comments by Vernon Smith in Cummings *et al.* 1984).

5. The method is well suited to measurement of nonuse benefits.

In summary, the current state of the art of the contingent valuation method had not advanced to the stage where one can categorically say that the method yields reasonably precise measures of social values. One can defensibly argue, however, that the methods generally yields order-of-magnitude estimates of values, and in some limited applications well-structured questionnaires may yield reasonably precise estimates (Cummings *et al.* 1984: Chapter 8).

Appropriate applications of the method

There seems to be general agreement – not to imply unanimity – that the CVM is most appropriate as a means for estimating values of public goods in the following circumstances:

The more familiar the subject is with the public good in question.

The less the degree of uncertainty associated with the public good (Kahneman and Tversky 1972, 1982b; Schoemaker and Kunreuther 1979; Schoemaker 1980; Just, Hueth, and Schmitz 1982).

When WTP measures as opposed to WTA measures are sought (Kahneman and Tversky 1972; Cummings *et al.* 1984).

The Contingent Ranking Method

The contingent ranking method (CRM), which is a variation of the above-described CVM approach, involves the use of surveys for acquiring data for estimating public goods benefits. The CRM is a relatively new method which has had a limited number of applications. It has been developed and applied to estimating the benefits of improved visibility by investigators at Charles River Associates. In this 1981 study subjects were given a set of cards each depicting a different situation with respect to visibility and other attributes of a national park, including a postulated admission fee. Respondents were asked to place their cards in order of preference. These rankings were then analyzed with a multinominal logit model which yields a set of parameter weights on the attributes which maximizes the likelihood of realizing that rank ordering. These parameter weights can then be used to determine the increase in the admission fee required to just offset the effect of an increase in visibility of the ranking of alternatives.

In contrast to the usual CVM approach, the CRM does not attempt to simulate a market per se. Rather, reflecting the random utility model (see Chapter 5), it requires that subjects order public-good payment combinations from most to least-preferred. From this ranking of alternatives, the researcher infers parameters for a random utility model from which the compensating surplus (analogous to the CVM's WTA) for changes in the provision level of the public good (or equivalent surplus for WTP to prevent it) is calculated. Procedural steps for applying the CRM are as follows.

1. Design of the survey instrument, which includes the following key features:
   a. Cogent explanation of the survey and its purposes.
   b. The CRM cards, each vividly describing a distinct provision level of the public good and the associated level of payment or compensation.
   c. Framing of the ranking question/request.
   d. Request for demographic and attitudinal data.
2. Pre-tests of the survey instrument.
3. Choice of sample design and survey area(s).
4. Training of surveyors.
5. Implementation of the survey.

Secondary data are not directly used in the CRM. For the sampling design, census data are often used for identifying households (or neighborhoods) to be included in the sample.

Output and interpretation

Data from the CRM are used to estimate scaled values for the random utility model. The resulting model may be used to define indifference curves and to estimate the change in subjects' income that would just offset utility gains (losses) from a change in the provision level of the public good. This estimated change in income corresponds to the measure of compensating surplus which, in turn, is the subject's maximum willingness to pay (accept) for the posited public good. This procedure assumes that WTP measures corresponding to changes in levels of the public good are the required measures of social benefits of interest in assessment of social policy. Alternatively, as discussed in Chapter 5, utilities may be assessed and then aggregated directly, without first encoding them in monetary terms.

A number of problems arise in structuring the random utility model including characterizing distribution functions for the random component of the model and specifying constraints on individual choice. A comprehensive review of these topics is found in a recent report prepared by the Environmental Protection Agency: Desvousages, Smith, and McGivney (1983).

Major assumptions

The following major assumptions underlie the contingent ranking method:

1. Individual behavior is represented by the random utility model.
2. There is an absence of "money illusion" on the part of subjects.
3. Irrelevant alternatives are not considered.

Strengths and weaknesses

The major weaknesses of the contingent ranking method are the following:

1. Uncertainty remains as to appropriate specifications of random utility models underlying the CRM.
2. Related to the above, when models are structured so as to reflect differences in individual determinants of utility, large (thus far nonquantified) numbers of rank alternatives are required. Research concerning (as examples) bounded rationality and cognitive dissonance raises serious questions as to subjects' incentives and capabilities to mentally process the large amounts of information implied by expanded rank alternatives.
3. The case has yet to be made that criticisms concerning hypothetical bias leveled at the CVM do not apply with equal force for the CRM. This may be particularly true for framing effects, issues related to hypothetical payment and attitude-behavior controversies. One may argue that unlike the CVM's reliance on market analogies, the CRM directly identifies preferences. But since resulting benefit estimates are interpreted within a context which implies behavior (derived benefit estimates are intended or assumed to reflect individual *actual* willingness to pay or accept), one must be assured that rankings imply something more than manifestations of attitudes.

Given the early stage of development of this method, its strengths must be viewed in terms of potential. Thus, for example, given resolution of the issues raised above, one claim to strengths for the CRM might be the fact that it is applicable to a wide range of public goods. Another is that, in contrast to the usual CVM approach, it requires only ordinal preference data, which are relatively easy to obtain and which respondents are often

6. General Methods for Benefits Assessment

relatively sure of. Additional claims to strengths for this method must await further research results.

Appropriate applications

Bearing in mind the caveats implied by the above discussions, applications of the CRM are likely to be most effective in situations where alternative provision levels of the public good are plausibly discrete and are viewed as such by subjects and in situations where direct values predominate values attributable to the public good.

The Petition Method

The petition method (PM), which again uses survey techniques, is not as yet as formally developed an estimation method as the contingent valuation method and, to a lesser extent, the contingent ranking method. Indeed, recent interest in the development of a PM has seemingly evolved as a result of problems encountered in the development of the CVM, primarily, problems associated with hypothetical payment within the context of a simulated market institution. Despite the formative stage of its development, the PM approach is considered here because of the strong possibility that the method will receive considerable attention in the benefits assessment literature in the near future. Procedural steps for applying the PM will take the following general form:

1. Design of the survey instrument, which includes these key features:
    a. Cogent explanation of the survey and its purposes, with stress given to the nationwide character of the survey and the potential influence of survey results on public policy; *signed* petitions call for action on the part of the government.
    b. Description of the petition-requested provision level of the public good.
    c. Description of additional costs which households must bear if the public good is provided. (This cost will vary among different subsets of subjects.)
    d. Question as to subjects' *willingness to sign* the petition.
    e. Request for demographic/attitudinal data.
2. Pre-tests of the survey instrument.
3. Choice of sample design and survey areas.
4. Training of surveyors.
5. Implementation of the survey.

Other than data used for structuring survey areas, secondary data may not be required for applications of the PM.

Output and interpretation

For the ith provision level of the public good, and the jth posited cost for the good, N_{ij} subjects are surveyed, from which n_{ij} subjects will have signed the petition. Regression and other statistical techniques may then be used to infer, for given demographic/attitudinal characteristics, willingness to pay for alternative levels of the public good. The resulting values are interpreted as the individuals' contingent willingness to pay for the posited change in the provision level of the public good.

Major assumptions

Given the early stage of development of the petition method, we know of no efforts designed to model individual behavior under PM conditions. Obvious assumptions underlying the method can be defined, however, as follows:

1. The act of signing the petition is indicative of contingent intended behavior as opposed to an indication of attitudinal preferences; that is, subjects do not sign petitions frivolously.

2. Subjects view as "real" the petition's potential effect on public policy *vis-à-vis* the provision of the public good and do not just sign it to please the canvasser, thinking that it will make no real difference.

Strengths and weaknesses

One of the major criticisms of the CVM has been that individuals are aware that decisions concerning the provision of public goods are not based on hypothetical statements of willingness to pay; such provisions typically result from political pressures (particularly in the case of environment-related "goods"). Therefore, exercises designed to elicit hypothetical WTP values may not be taken seriously. Herein lies the major potential strength of the PM, which does not rely on a hypothetical market institution; rather, it relies on an actual *political* petition wherein petition signers urge government action. The PM purports to provide an institution related to the contingent provision of a public good which closely squares with individual experiences and perceptions. Of course, at this point in time such strengths are little more than conjectures.

Absent theoretical and empirical development of the PM, one can anticipate several potential weaknesses. First, agency reactions to a method where surveyors systematically "lie" to subjects (regarding, e.g., the nationwide survey costs and the provision of the public good) is problematic. Secondly, notwithstanding the arguable appeal of the petition institution *vis-à-vis* the market institution, immediately apparent means do not exist by which the accuracy of measures derived with the PM can be assessed.

Appropriate applications

All else equal, the PM may be applicable to a wide range of public goods. The obvious limitation of the method is that it must be used only for those goods whose provision might credibly be influenced by a petition. Thus, environmental improvements might be valued with the PM but not public goods such as national defense.

Indirect Monetary Methods for Evaluating Public Goods

We now turn from direct (survey-based) to indirect (market-based) methods for evaluating public goods.

The Hedonic Price Method

The hedonic price method (HPM) for valuing public, nonmarket goods involves efforts to impute public good values from observed market prices (Ronan 1973; Rosen 1974; Freeman 1979; Cummings, Schulze, and Brookshire 1982; Brookshire et al. 1982b; Crocker and Forster 1984). If the level of a nonmarket good, such as safety, noncongestion, or air quality varies geographically, then the level of the good at different locations may be reflected in, for example, local housing prices, which therefore contain information about the implicit market value of the good. The hedonic price method attempts to extract such value information from the geographic variation in housing prices as a function of housing attributes, including nonmarket ones. Suppose that a product or commodity which commands a market price P has attributes $a_1, \ldots, a_n$, where attribute a_n is the public good in question. As an example, a house will have neighborhood-specific attributes such as crime rates, school quality, and air quality. Suppose our interest is in the public good – air quality. Regression techniques are then used to impute that part of the changes in market (housing) prices P that are attributable to changes in the air quality attribute. Procedural steps for applying the HPM are as follows (see Brookshire et al. 1982b):

1. Define the market commodity which includes as one of its attributes the public good in question; specify the functional relationship between the market prices and all relevant attributes of the market commodity.

2. Collect cross-sectional and, when appropriate, time-series data on market prices for the commodity and associated attribute values.

3. Conduct statistical analyses.

These steps and the underlying theory of the HPM are described more fully below, where the example of the housing market is explored in more detail along with other applications of the method.

Data requirements

Large amounts of data are often required for applications of the HPM. Continuing the property value–air quality example, for several communities with distinct differences in air quality, one must acquire, in addition to air quality measures, real estate sales data, characteristics of the houses sold (e.g., house size, number of bedrooms), and neighborhood characteristics (e.g., school quality, crime rates) relevant for houses sold (Brookshire et al. 1982b; Schulze et al. 1983a). Unintentional omission of relevant attributes (explanatory variables) can confound the interpretation of market data in terms of implicit prices.

Output and interpretation

The output of the HPM is a price function, or rent gradient (see Brookshire et al. 1982b: 167–169), which relates value and the level of provision of the public good. Under the best of circumstances – when included variables adequately account for differences in tastes, income, and all relevant characteristics of the public good and when price adequately reflects value (i.e., when all surplus is distributed in the market) – the price function derived with the HPM provides the desired measure of social benefits attributable to a given provision level of the public good. Under circumstances less than best, the price function desired from the HPM may generally provide value estimates (i.e., marginal implicit prices) that are lower bounds on maximum willingness to pay for the public good. The potential underestimation of maximum willingness to pay by the HPM's price function results from the effects of individual budget constraints on an individual's ability to trade off, in an optimal fashion, utility maximizing levels of the public good with other goods (income).

Major assumptions

Aside from standard assumptions concerning individual behavior received from the economist's theory of value, applications of the HPM require one major assumption: in making value decisions, individuals are cognizant of and optimize across the attributes of market commodities; that is, market commodities are differentiated by their attributes in ways consonant with the effective working of an implicit market for such attributes.

Strengths and weaknesses

The major strength of the HPM is that, *ceteris paribus,* it yields the market equilibrium value of the public good. This strength, however, is subject to the following caveats which reflect the method's weaknesses (Brookshire *et al.* 1982b):

1. Aside from ever-present problems of model specification, including

choice of functional form and explanatory variables (attributes) to be included, estimation of even a single equation price function may be, and often is, hindered by severe empirical difficulties, primarily multicollinearity and uncertainty about the appropriate functional form (Maler 1977).

2. Serious questions exist as to the extent to which the actual market equilibrium matching of buyers and sellers leads to a continuous price function and, *vis-à-vis* the implicit market of the HPM (V. K. Smith 1984: 23–24,) the extent to which individual *perceptions* of commodity attributes are consonant with the assumptions of the HPM (Burt and Brewer, 1971; Vaughn and Russell 1982). For example, public goods such as small, widely spread improvements in water quality may not be perceivable by individuals.

3. It is not clear that the value imputed to a public good by the HPM from study of one market – for example, property values for houses – captures the *full* value of the public good. Clean air, for example, affects people at work as well as at home. Therefore, wage differentials may have to be considered as well as property value differentials to get a full picture of the value from cleaner air.

4. Little is known concerning appropriate confidence intervals for HPM measures *vis-à-vis* "accurate" measures of social value for public goods.

5. As implied above, data acquisition may be quite expensive, or adequate data may simply not be available.

6. The method has limited applications.

Appropriate applications

In theory, the HPM might be appropriately applied to the valuation of any public good. In fact, the substance of the method imposes severe restrictions on its appropriate domain of application. What is required is the existence of a private commodity traded in the marketplace which has as one of its dominant attributes the public good of interest to the researcher, so that "weak complementarity" can be used to establish a link between the price of the market good and the price imputed to the public good. The strategies of using weak complementarity, as well as substitute and complement effects between a market and a nonmarket good (such as environmental quality) to derive an implicit price for the nonmarket good, are described and discussed in Freeman (1979b).

Example: Hedonic price method and the housing market

The HPM is designed to estimate implicit prices of the characteristics which differentiate closely related products in a product class. For example, houses constitute a product class differentiated by characteristics such as the number of rooms, size of lot, and location. If there is sufficient diversity among houses on the market, then in principle it is possible to estimate the implicit price relationship which gives the price of each type of house as a function of the levels of its various attributes. The implicit prices of characteristics are given by the partial derivatives of the hedonic price equation with respect to these characteristics.

More formally, assume that individual utility is a function of housing services consumed as well as of the consumption bundle of other market goods X. Housing services are defined by the levels of the attributes of the housing bundle consumed. Let the attributes be represented by S, N, and Q, where S is a vector of structural attributes such as size and number of rooms, N is a vector of neighborhood attributes such as accessibility and quality of schools, and Q is a measure of, for example, air quality at the house site. The price of any house will depend upon the quantities of the various attributes that it embodies. Let $P_H = P_H(S,N,Q)$ be the function that maps bundles of attribute levels to market prices of houses. This is the *hedonic price function*. The partial derivative of this function with respect to any attribute, say Q, is the *marginal implicit price* of that attribute. In general, the marginal implicit price of any attribute can be a function of all the other attributes of housing.

Assuming that the individual consumes only one house, his choice problem is to

maximize $U(X,S,N,Q)$ subject to $P \cdot X + P_H = M$

where P is a vector of market prices and M is his wealth constraint.

If the choice set of houses (attribute combinations) that he must choose from is so rich that $P_H(\ )$ can be approximated as a differentiable function and the choice set can be modelled as a continuous set, then under various technical assumptions (e.g., that the choice set is closed, convex, and bounded) the first-order conditions for solution of his choice problem include

$$\frac{\partial P_H}{\partial Q} = P_{X_i} \cdot MRS_{QX_i}$$

The right-hand side of this condition is his marginal willingness to pay for Q. The left-hand side is the marginal implicit price of Q reflected in the market, as revealed by the hedonic price function. In other words, if the individual is in equilibrium, the marginal implicit price of Q for that individual can be taken as a measure of the individual's marginal willingness to pay for an improvement in Q.

If the hedonic price function is known, the marginal implicit prices of any attribute Q, such as air quality or safety of the neighborhood, can be computed for each individual in the housing market. The sum of these individual marginal implicit prices may then be defined as the aggregate marginal benefit for a change in Q affecting all houses in the market.

Nonmarginal changes and aggregation

In order to calculate each individual's benefit from a *nonmarginal* change in Q, knowledge of each individual's inverse demand (or marginal WTP) function for Q is required. Where this information is not available, several approaches for approximating individual benefits have been identified (see Freeman 1979a).

If the hedonic price function is nonlinear (and there is no reason to expect linearity in the case of housing), then *different individuals will in general have different marginal implicit prices* and values for Q and other attributes making it difficult to identify a meaningul inverse aggregate demand curve. (An exception is if all individuals have identical incomes and utility functions: then the marginal implicit price function is itself the inverse demand function.) Where differences in incomes, preferences, or other variables lead to different inverse demand functions, however, recovery of the inverse demand functions for individuals is not straightforward. See Brown and Rosen (1982: 328–329).

The problem is that data from a single market reveal only the outcome of a single market experiment. Since each individual faces a nonlinear budget constraint, he must choose the marginal value and the quantity of Q simultaneously. Different value/quantity combinations chosen by individuals reflect differences in incomes, preferences, and other determinants of demand. The data available from a single hedonic market do not provide a basis for sorting out these other influences on the value/quantity choice.

In principle, there are two possible approaches to resolving this problem. The first is to increase the quantity of information obtained from marginal implicit prices by estimating hedonic price functions for several separate markets and then pooling the cross-sectional data on the assumption that the underlying structure of supply and demand is the same in all markets (Freeman 1979a; Brookshire *et al.* 1984). The second approach is to impose additional structure on the problem by invoking an a priori assumption about the form of the underlying utility function. Quigley (1982) has shown how estimates of marginal implicit prices can be used to estimate the parameters of the previously specified generalized constant elasticity of substitution utility function. If the utility function correctly describes preferences, then once its parameters are known, estimating the benefits of changes in attributes is straightforward.

In summary, the hedonic price model provides an attractive approach to estimating demands for site amenities such as air quality, absence of noise, and scenic views. But both the estimation and interpretation of hedonic price functions are fraught with more problems than early advocates of this approach recognized. Data problems and misspecification problems in the estimation stage can lead to errors in the measurement of marginal prices for individuals. And the practical approaches to approximating inverse demand functions and the benefits of nonmarginal changes in Q introduce further errors in aggregate benefits estimates. Nevertheless, despite these problems, the hedonic price model does provide a basis for reducing our ignorance about the magnitude of the benefits associated with changes in site-specific amenities.

Urban amenities and hedonic wages

From a worker's perspective, a job can be viewed as a differentiated product, that is, a good with a bundle of characteristics such as working conditions, prestige, training and enhancement of skills, and degree of risk of accidental injury or exposure to toxic substances. If workers are free to move from one urban area to another, then jobs are differentiated in part by the environmental and other characteristics of the urban area where the job is located. If workers can be assumed to be free to choose from a sufficiently rich menu of differentiated jobs, then the HPM can be applied to data on wages, job safety, and other job attributes to try to estimate the marginal implicit prices of these attributes. In hedonic wage studies, the employer is viewed as selling a package of job attributes, but also as purchasing work effort. Thus, the hedonic wage equation must be interpreted as a reduced form equation reflecting not only the interaction of supply and demand for job attributes, as in the housing market example, but also the interaction of supply and demand for worker characteristics (Lucas 1977; Rosen 1974). This means that both workers and job characteristics must be included as arguments in the estimated hedonic wage equation.

When hedonic wage or price functions are estimated, it is necessary that the labor market be in equilibrium *and* that the labor market *not* be segmented into submarkets or regions with incomplete mobility among segments (Freeman 1979a, 1979b). When hedonic wage equations are estimated using data from several urban areas, it is necessary to assume that these areas are part of a single market. In general, the extent of market segmentation and its significance for empirical estimation of hedonic wage functions are not known.

The Travel Cost Method

The travel cost method (TCM) is used to estimate individual demands for inputs to a household production function – typically the household recreation production function (see Mendelsohn 1980; Brown and Rosen 1982; and Bishop, Heberlein, and Kealy 1983). Usually benefits attributable to a recreation site are obtained by integrating under this derived Marshallian demand curve, thereby getting a measure of consumer surplus. Procedural steps for applying the TCM are as follows (see Brown and Rosen 1982):

6. General Methods for Benefits Assessment

1. Prepare TCM questionnaire, which includes the following components:
   a. Place of residence and other demographic/attitudinal information.
   b. Frequency of visits to site in question.
   c. Frequency of visits to substitute sites.
   f. Trip information: purposes of trip other than visitation of the site in question, length of trip, nights stayed in motels, travel paths, meals at restaurants (trip cost information).
2. Implement survey (see McConnell and Bockstael 1984).
3. Estimate travel costs for individuals or for distance zones from the site.
4. Develop measures of characteristics for the site in question and for substitute ones.
5. Develop system of demand equations to be used for the TCM analysis (see Burt and Brewer 1971; Vaughn and Russell 1982).

Secondary data can be used for procedural steps 3 and 4. For step 1, primary data are acquired via the questionnaire, which is applied at entrances to the recreation site or, in some cases, by a mail survey (Bishop and Heberlein 1979).

Output and interpretation

As noted above, the TCM yields estimates for the demand curve relevant for a particular recreation site. Net benefit measures derived from the TCM must be interpreted with the following consideration in mind. Benefit estimates derived from the TCM apply to the site *per se*, not to the joint outputs of the site (e.g., specific recreation experiences) (Bockstael and McConnell 1981). The method says nothing about the value of any part of the site or the value of altering characteristics of the site.[2] Given that trip-related, on-site utility gains are typically ignored, the cost-based TCM estimate of benefits may generally be considered as underestimating social benefits from the site.[3]

Major assumptions

Major assumptions underlying general applications of the TCM include the following (McConnell and Bockstael 1984):

1. Individuals take trips to the point where marginal trip costs equal marginal site values, and they react to an entrance fee in the same manner as to an increase in travel cost.
2. There is homogeneity of on-site outputs of trips across site visitors.

Given the various possible demand structures which one might use in applying the TCM (see Mendelsohn and Brown 1983), other assumptions may be relevant for a specific application. Such assumptions are implied in the description of strengths and weaknesses.

3. Individuals take trips for the sole purpose of visiting the site.

Strengths and weaknesses

The major strengths of the TCM lie, first, in its focus on observable purchase of inputs, whereby one avoids the many identification problems encountered with other (household production function) approaches. Secondly, for somewhat limited policy purposes, simple applications of the TCM can be used to generate relatively inexpensive order-of-magnitude estimates for single-site benefits.

Major weaknesses of the TCM are as follows:

1. The general lack of distinction between the demand for trips to a site and on-site utility gains (although this weakness can be alleviated using logit models of demand).

2. Difficulties in valuing or costing travel time to the site, especially for very short trips.

3. Difficulties in adjusting costs to reflect multi-purpose trips.

4. Substantial data requirements for robust applications which include relevant price-quantity-characteristic representations for alternative substitute sites. It must be assumed that analysts and respondents use the same mental model of site attributes and values.

Appropriate applications

The use of the TCM is obviously limited to a particular type of nonmarket good – visited sites. Taking into consideration data acquisition costs and likely availability of data, as well as difficulties encountered in data analyses, the method works best when applied to the valuation of a single site (with unchanged characteristics) for which conditions at alternative sites can be expected to remain constant. For more complex problems, the method works best the fewer the number of site changes and alternative sites available to individuals.

Other Indirect Valuation Methods

The hedonic price and travel cost methods described above are the primary indirect valuation methods used by researchers in current efforts to estimate social benefits attributable to public goods. For completeness, it is desirable to include in this section a brief sketch of two additional methods

6. General Methods for Benefits Assessment

for valuing public goods which have evolved from research using the TCM: the site substitution method and the household production function method. Here we eschew detail, given their similarity – in terms of data requirements, appropriate uses, and so forth – to the TCM.

The site substitution method

The site substitution method (SSM) is sometimes used as a part of the TCM when changes in site characteristics are to be valued in the study (see Knetsch and Davis 1966 and Thayer 1981). The essence of the SSM is quite simple. One applies the TCM as described above, with one question added to the questionnaire: given (environmental change at the site in question), would this change affect your visitation to this site, and if so, what alternative site would you visit as a result of this change? The (additional) travel costs required for visitations to the substitute site are then interpreted as a very crude monetary measure of foregone utilities/benefits attributable to the posited, site-specific environmental change, assuming that the sites are otherwise comparable. This approach is perhaps the weakest of those surveyed, as adjustments for different utility levels at different sites cannot easily be made.

The household production function method

The household production function method (HPFM) focuses on commodities produced by the household using as input purchased commodities and time. In contrast to the TCM, which measures the value of acquiring the site *per se,* the HPFM values the outputs (driving, recreation experiences, etc.) produced at the site. Households are assumed to derive utility from commodities X produced by the household, where some X include recreation experiences. One assumes a specific form for the production function used by households in producing these commodities, where inputs to the production function include goods purchased as well as nonmarket goods such as cleanliness of recreational waters, or markets and household time. Such inputs imply a cost function for household production which includes exogenous prices for purchased goods and the marginal wage rate. Familiar production economics techniques (Bockstael and McConnell 1981) are used to derive demand functions using the above-described functional relationships.

The HPFM works best when all inputs in the production function are purchased in markets and when production technologies are linear; its appeal *vis-à-vis* the TCM is in its focus on utility gains from total recreation experiences. The method has severe weaknesses, however. First, the linear technology assumption is strong, and the use of nonlinear technologies requires extraordinarily large data sets (Mendelsohn and Brown 1983). Similar problems arise with the equally strong assumption of homogeneous (across households) production functions commonly used in the HPFM. An exacerbating problem is the need for data information concerning *all* inputs and outputs of the household. At this point in time, data information requirements of the HPFM may justify Mendelsohn and Brown's conclusion that the tool is an unnecessarily cumbersome approach to measure the value of sites or their qualities (1983:611).

Summary

We have surveyed several widely used and some recently developed or developing methodologies for assessing the benefits of nonmarket goods, including the public and open-access goods associated with many health, safety, and environmental regulations. These methods build on the more familiar tools, such as producer and consumer surplus, and so forth associated with benefits assessment for market goods (see e.g., Freeman 1979b:330), but attempt to extend them to goods for which no well-defined markets exist.

Our review has focused on five methods: the contingent valuation, contingent ranking, petition, hedonic price, and travel cost methods. Other related approaches, such as voting and referendum methods other than the petition method, and the household production function method as an alternative to the travel cost method, have also been mentioned.

To help the practitioner compare and choose among these approaches, we have summarized some of their key properties in Table 6-1. The comments on data requirements, assumptions, and interpretation of output apply not only to the specific methods surveyed but also to related

Table 6–1. Summary of benefits assessment methods

Method	Data requirements	Assumptions	Output	Principal application areas
Contingent valuation method (CVM)	Survey sample data	Coherent preferences, truthfully expressed	"Simulated market" demand curve	Any public good for which people have well-defined holistic preferences
	Cardinal preference data (e.g., WTP/WTA)	WTP/WTA values can be aggregated meaningfully		
Contingent ranking method (CRM)	Survey sample data	Random utility preference structure	Estimated utility functions	Public goods with discrete alternative levels
	Ordinal preference data (e.g., ranking)	Behaviorally accurate responses		

6. General Methods for Benefits Assessment

Table 6-1 (continued)

Method	Data requirements	Assumptions	Output	Principal application areas
Petition method (PM)	Survey sample data	Responses reflect intended behavior	Data for logistic or other regression model	Public goods for which petitions are credible
	Binary choice data			
Hedonic price method (HPM)	Extensive market price data on diverse set of attribute combinations	Full value of public good is expressed through revealed preferences	Implicit prices for attributes	Small changes in environmental quality
		Assumed utility function form for non-marginal changes		
Travel cost method (TCM)	Secondary data on travel costs, site attributes	Sole-purpose site visits	Site-specific demand curve	Recreational sites
	Primary data on visitor characteristics	Marginal trip cost = marginal site benefit		

methods. For example, the comments on the HPM apply to econometric approaches generally, while those on the PM apply to related techniques such as voting and referendum methods.

We have not tried to be exhaustive in our summary of methods, since there have been many suggested approaches that do not seem to be practical or sound. In addition, as mentioned at the outset, we have in this chapter adopted the convention that all benefits are to be assessed in monetary terms, since this is usually done, even though survey-based approaches are quite compatible with the social utility approach described in Chapter 5. However, the five methods in Table 6-1 fairly well represent the range of available techniques in the current state of the art and indicate both their strengths and their limitations.

University of New Mexico

Arthur D. Little, Inc.

Bowdoin College

Notes

1. The PM and related approaches are topics of current research. Robert Mitchell, at Resources for the Future, is studying a similar political participation mechanism – voting in a referendum – within an overall CVM framework. The PM is also implied as a plausible, alternative institution for the CVM in Randall (n.d.). For completed studies using methods suggestive of the PM, see Seller, Stoll, and Chaves (n.d.), and Strand (n.d.).

2. A method for using the TCM for valuing such alternatives is suggested in the work of Vaughn and Russell (1982); such applications require substantial data concerning site characteristics (see 1982: 226, 227).

3. As noted below, other characteristics of the TCM may lead to overestimates, however, particularly the assumption of single-purpose trips.

References and Bibliography

Ajzen, I. and M. Fishbein. 1977. 'Attitude-Behaviors Relations: A Theoretical Analysis and Review of Empirical Research.' *Psychological Bulletin* 84: 888–918.

Akerlof, G. A. and W. T. Dickens. 1982. 'The Economic Consequences of Cognitive Dissonance.' *American Economic Review* 72(3).

Arrow, K. J. 1951. *Social Choice and Individual Values*. New York: Wiley. 1982.

———. 'Risk Perceptions in Psychology and Economics.' *Economic Inquiry* 20: 1–9.

Batie, S. S. and L. Shabman. 1979. 'Valuing Non-Market Goods – Conceptual and Empirical Issues: Discussion.' *American Journal of Agricultural Economics* 61(5): 931–932.

Belovicz, M. W. 1979. 'Sealed-Bid Auctions: Experimental Results and Applications.' In V. L. Smith, ed., *Research in Experimental Economics*. Vol. I. JAI Press.

Bergson, A. 1938. 'A Reformation of Certain Aspects of Welfare Economics.' *Quarterly Journal of Economics*. Feb.

Bishop, R. C. and T. A. Heberlein. 1979. 'Measuring Values of Extra-Market Goods: Are Indirect Measures Biased?' *American Journal of Agricultural Economics* 61: 926–930.

———, ———, and M. J. Kealy. 1983. 'Contingent Valuation of Environmental Assets: Comparison with a Simulated Market.' *Natural Resources Journal* 23(3): 619–634.

Blank, F. M., D. S. Brookshire, T. D. Crocker, R. C. d'Arge, R. L Horst, and R. D. Rowe. 1978. *Valuation of Aesthetic Preferences: A Case Study of the Economic Value of Visibility*. Prepared for the Electric Power Research Institute, Palo Alto.

Bockstael, N. and K. E. McConnell. 1981. 'Theory and Estimation of the Household Production Function for Wildlife Recreation.' *Journal of Environmental Economics and Management* 8:199.

———, ———. 1983. 'Welfare Measurement in the Household Production Function.' *American Economic Review* 73(4): 806–814.

Bohm, P. 1971. 'An Approach to the Problem of Estimating Demand for Public Goods.' *Swedish Journal of Economics* 73(1): 55–66.

———. 1972. 'Estimating Demand for Public Goods: An Experiment.' *European Economic Review* 3(2): 111–130.

Boulding, K. E. 1975. *The Image*. Ann Arbor: University of Michigan Press.

Bradford, D. F. 1970. 'Benefit-Cost Analysis and Demand Curves for Public Goods.' *Kyklos* 23: 775–791.

Brookshire, D. S. and R. C. d'Arge. 1979. 'Resource Impacted Communities: Economics, Planning, and Management.' Paper prepared for the 4th U.S.-U.S.S.R. Symposium on Comprehensive Analysis of the Environment. Jackson, Wyoming.

———, ———, W. D. Schulze, and M. A. Thayer. 1981. 'Experiments in Valuing Public Goods.' In V. K. Smith, ed., *Advances in Applied Microeconomics*, Vol. I. JAI Press.

——, R. G. Cummings, M. Rahmatian, W. D. Schulze, and M. A. Thayer. 1982. 'Experimental Approaches for Valuing Environmental Commodities.' *Methods Development in Measuring Benefits of Environmental Improvements*. Draft final report to the Environmental Protection Agency.

—— and T. Crocker 1981. 'The Advantages of Contingent Valuation Methods for Benefit-Cost Analysis.' *Public Choice* 36(2): 235–252.

—— and L. S. Eubanks. 1978. 'Contingent Valuation and Revealing the Actual Demand for Public Environmental Commodities.' In *Public Choice in New Orleans*, G. Tullock, ed., selected paper of the Public Choice Society, New Orleans.

——, ——, and A. Randall. 1983. 'Estimating Option Prices and Existence Values for Wildlife Resources.' *Land Economics* 59: 1–5.

——, B. C. Ives, and W. D. Schulze. 1976. 'The Valuation of Aesthetic Preferences.' *Journal of Environmental Economics and Management* 3: 325–346.

——, A. Randall, R. C. d'Arge, L. S. Eubanks, J. R. Stoll, T. D. Crocker, and S. Johnson. 1977. *Methodological Experiments in Valuing Wildlife Resources: Phase I Interim Report to the United States Fish and Wildlife Service*.

——, ——, and J. R. Stoll. 1980. 'Valuing Increments and Decrements in Natural Resource Service Flows.' *American Journal of Agricultural Economics* 62(3): 478–488.

——, W. D. Schulze, M. A. Thayer, and R. C. d'Arge. 1982. 'Valuing Public Goods: A Comparison of Survey and Hedonic Approaches.' *American Economic Review* 72(1): 165–177.

——, ——, and R. Whitworth. 1983. *An Economic Analysis of the Private Sector Benefits and Potential for Cost Recovery of the National Coal Resources Data Systems*. Report to the U.S. Geological Survey, Reston, Va.

——, M. S. Thayer, J. Tschirhart, and W. D. Schulze. 1984. 'A Test of the Expected Utility Model: Evidence from Earthquake Risks.' Conditional acceptance in *Journal of Political Economy*.

Brown, J. and H. Rosen. 1982. 'On Estimation of Structural Hedonic Price Models.' *Econometrica* 50(3): 765–768.

Brown, W., A. Singh, and E. Castle. 1964. *An Economic Evaluation of the Oregon Salmon and Steelhead Sports Fishery*. Oregon Agricultural Experiment Station Bulletin, No. 74, Corvallis.

Buchanan, J. and G. Tullock. 1962. *The Calculus of Consent*. Ann Arbor: University of Michigan Press.

Burness, H. S., R. G. Cummings, A. F. Mehr, and M. S. Walbert. 1983. 'Valuing Policies which Reduce Environmental Risk.' *Natural Resources Journal* 23(3): 675–682.

Burt O. and M. Brewer. 1971. 'Estimation of Net Social Benefits from Outdoor Recreation.' *Econometrica* 39: 813.

Carson, R. T. and R. C. Mitchell. 1984. 'A Reestimation of Bishop and Heberlein's Simulated Market-Hypothetical Markets-Travel Cost Results under Alternative Assumptions.' Forthcoming. *American Journal of Agricultural Economics*.

Cassady, R. 1967. *Auctions and Auctioneering*. Berkeley: University of California Press.

Cesario, F. J. 1976. 'Valuing Time in Recreation Studies.' *Land Economics:* 52.

Chew, S. H. and K. R. MacCrimmon. 1979. 'Alpha Utility Theory, Lottery Composition and the Allais Paradox.' Working paper, Faculty of Commerce and Business Administration, University of British Columbia.

Ciriacy-Wantrup, S.V. 1952. *Resource Conservation: Economics and Policies*. Berkeley: University of California Press.

Clarke, E. H. 1971. 'Multipart Pricing of Public Goods.' *Public Choice* 11: 17–33.

Clawson, M. 1959. 'Methods of Measuring the Demand for Outdoor Recreation.' Reprint No. 10. Washington D. C.: Resources for the Future.

—— and J. Knetsch. 1966. *Economics of Outdoor Recreation*. Baltimore: Johns Hopkins University Press.

Coppinger, V. M., V. L. Smith, and J. A. Titus. 1980. 'Incentives and Behavior in English,

Dutch and Sealed-Bid Auctions.' *Economic Inquiry* 18(1): 1–22.
Coursey, D. L. and V. L. Smith. 1982. 'Experimental Tests of an Allocation Mechanism for Private, Public, or Externality Goods.' Mimeographed manuscript, Department of Economics, University of Arizona.
—— and H. Nyquist. 1983. 'Application of Robust Estimation Techniques in Demand Analysis.' Manuscript, University of Umea, Sweden, submitted to *Review of Economics and Statistics*.
Cox, J. L., B. Roberson, and V. L. Smith. 1982. 'Theory and Behavior of Single Price Auctions.' In V. L. Smith, ed., *Research in Experimental Economics*. Vol. II. JAI Press.
Crocker, T. D. 1984. 'On the Value of the Condition of a Forest Stock.' Unpublished manuscript, Department of Economics, University of Wyoming.
—— and R. G. Cummings. Forthcoming. 'On Valuing Deposition-Induced Materials Damages: A Methodolgical Inquiry. In D. D. Adams and W. Page, eds., *Acid Deposition: Environmental and Economic Impacts*. New York: Plenum Press.
—— and B. A. Forster. Forthcoming. 'Nonconvexities in the Control of Acid Deposition: Still a Problem.' *Journal of the Air Pollution Control Association*.
Cummings, R. G., H. S. Burness, and R. D. Norton. 1981. 'Measuring Household Soiling Damages from Suspended Air Particulates: A Methodological Inquiry.' Report to the Environmental Protection Agency.
——, W. D. Schulze, and A. J. Mehr. 1978. 'Optimal Municipal Investment in Boomtowns: An Empirical Analysis.' *Journal of Environmental Economics and Management* 5 (Sept.).
——, ——, and D. S. Brookshire. 1983. 'Measuring the Elasticity of Substitution of Wages for Municipal Infrastructure: A Comparison of the Survey and Wage Hedonic Approach.' Unpublished manuscript.
—— et al. 1984. 'Valuing Public Goods: A State of the Arts Assessment of the Contingent Valuation Method.' Manuscript.
Dasgupta, A. K. and D. W. Pearce. 1978. *Cost-Benefit Analysis: Theory and Practice*. New York: Macmillan.
Davis, O. and A. Whinston. 1966. 'Welfare Economics and the Theory of the Second Best.' *Review of Economic Studies*.
Davis, R. K. 1963a. 'Recreational Planning as an Economic Problem.' *Natural Resource Journal* 3. 238–249.
——. 1963b. 'The Value of Outdoor Recreation: An Economic Study of the Maine Woods.' Unpublished Ph.D. dissertation, Harvard University.
Daubert, J. and R. Young. 1981. 'Recreational Demands for Maintaining Instream Flows: A Contingent Valuation Approach.' *American Journal of Agricultural Economics* 63(4): 666–676.
Desvousges, W. H. and V. K. Smith. 1982. *The Basis for Measuring the Benefits of Hazardous Waste Disposal Regulation*. Raleigh, N. C.: Research Triangle Institute.
——, ——, and M. P. McGivney. 1983. *A Comparison of Alternative Approaches for Estimating Recreation and Related Benefits of Water Quality Improvements*. EPA–230–05–83–001.
Devletoglou, N. E. 1971. 'Threshholds and Transaction Costs.' *Quarterly Journal of Economics* 85(1).
Dubey, P. and M. Shubik. 1980. 'A Strategic Market Game with Price and Quantity Strategies.' *Zeitschrift für Nationalökonomie* 40.
Engelbrecht-Wiggans, R. 1980. 'Auctions and Bidding Models: A Survey.' *Management Science* 26 (2):119–142.
Feather, T. 1959. 'Subjective Probability and Decisions under Uncertainty.' *Psychology Review* 66: 150–164.
Feenburg, D. and E. S. Mills. 1980. *Measuring the Benefits of Water Pollution Abatement*. New York: Academic Press.
Ferejohn, J. A., R. Forsythe, and R. Noll. 1979a. 'An Experimental Analysis of Decision-

making Procedures for Discrete Public Goods: A Case Study of a Problem of Institutional Design.' In V. L. Smith, ed., *Research in Experimental Economics*. Vol. I. JAI Press.

———, ———, and ———. 1979b. 'Practical Aspects of the Construction of Decentralized Decisionmaking Systems for Public Goods.' In C. Russell, ed., *Collective Decision Making: Applications from public Choice Theory*. Baltimore: Johns Hopkins University Press.

———, ———, ———, and T. R. Palfrey. 1982. 'An Experimental Examination of Auction Mechanisms for Discrete Public Goods.' In V. L. Smith, ed., *Research in Experimental Economics*. Vol. I. JAI Press.

——— and R. G. Noll. 1976. 'An Experimental Market for Public Goods: The PBS Station Program Cooperative.' *American Economic Review Proceedings* 66: 267–273.

Fischhoff, B. 1975. 'Hindsight is not Equal to Foresight: The Effect of Outcome Knowledge on Judgment under Uncertainty.' *Journal of Experimental Psychology* 104(1): 288–299.

———. 1982. 'For Those Condemned to Study the Past: Heuristics and Biases in Hindsight.' In *Judgment under Uncertainty: Heuristics and Biases*. D. Kahneman, P. Slovic, and A. Tversky, eds. Cambridge: Cambridge University Press.

———, P. Slovic and S. Lichtenstein. 1982. 'Lay Foibles and Expert Fables in Judgments about Risk.' *The American Statistician* 36(2): 240–255.

Forsythe. R. and R. M. Isaac. 1982. 'Demand-Revealing Mechanisms for Private Goods Auctions.' In V. L. Smith, ed., *Research in Experimental Economics*, Vol. II. JAI Press.

Freeman, A. M. III. 1979a. 'Approaches to Measuring Public Goods Demands.' *American Journal of Agricultural Economics Proceedings* 61(5): 915–920.

———. 1979b. *The Benefits of Environmental Improvement*. Baltimore: Johns Hopkins University Press.

Friedman, M. 1953. *Essays in Positive Economics*. Chicago: University of Chicago Press.

Gallagher, D. R. and V. K. Smith. Forthcoming. 'Measuring Values for Environmental Resources under Uncertainty.' *Journal of Environmental Economics and Management*.

Georgescu-Roegen, N. 1958. 'Threshhold in Choice and the Theory of Demand.' *Econometrica* 26: 157–168.

Gerking, S. and W. D. Schulze. 1981. 'What Do We Know about Benefits of Reduced Mortality from Air Pollution Control.' *American Economic Review* 71(2): 228–240.

Gordon, I. M. and J. L. Knetsch. 1979. 'Consumer's Surplus Measures and the Evaluation of Resources.' *Land Economics* 55: 1–10.

Gramlich, F. 1977. 'The Demand for Clean Water: The Case of the Charles River.' *National Tax Journal* 30(2): 183–194.

Greenley, D. A., R. Walsh, and R. A. Young. 1981. 'Option Value: Empirical Evidence from a Case Study of Recreation and Water Quality.' *Quarterly Journal of Economics* 86: 657–673.

———, ———, and ———. 1982. *Economic Benefits of Improved Water Quality*. Boulder: Westview.

Gregory, R. S. 1982. 'Valuing Non-Market Goods: An Analysis of Alternative Approaches.' Unpublished Ph.D. dissertation. University of British Columbia.

Grether, D. M. and C. R. Plott. 1979. 'Economic Theory of Choice and the Preference Reversal Phenomenon.' *American Economic Review* 69: 623–638.

Groves, T. 1973. 'Incentives in Teams.' *Econometrica* 41: 617–631.

——— and J. Ledyard. 1977. 'Optimal Allocation of Public Goods: A Solution to the "Free Rider Problem." ' *Econometrica* 45(4): 763–809.

Hershey, J. C. and P. J. H. Schoemaker. 1980. 'Risk Taking and Problem Context in the Domain of Losses: An Expected Utility Analysis.' *Journal of Risk and Insurance* 47: 111–132.

Hey, J. D. 1983. 'Whither Uncertainty.' *Economic Journal* (supplement): 130–139.

Heyne, P. 1983. *The Economic Way of Thinking*. 4th ed. Chicago: Science Research Associates.

Hoehn, J. P. and A. Randall. n.d. 'Aggregation and Disaggregation of Program Benefits in a Complex Policy Environment.' Unpublished manuscript.

———, ———. 1983. 'Incentives and Performance in Contingent Valuation.' Mimeographed manuscript. University of Kentucky.
Hori, H. 1975. 'Revealed Preference for Public Goods.' *American Economic Review* 65.
Hovis, J., D. C. Coursey, and W. D. Schulze. 1983. 'A Comparison of Alternative Valuation Mechanisms for Non-Market Commodities.' Unpublished manuscript. Department of Economics, University of Wyoming.
Hurwicz, L. 1973. 'The Design of Mechanisms for Resource Allocation.' *American Economic Review Proceedings*.
Isaac, R. M. 1983. 'Laboratory Experimental Economics as a Tool in Public Policy Analysis.' Mimeographed manuscript. Department of Economics, University of Arizona.
Jarecki, H. 1976. 'Bullion Dealing, Commodity Exchange Trading, and the London Gold Fixing: Three Forms of Commodity Auctions.' In Y. Amihud, ed., *Bidding and Auctioning for Procurement and Allocation*. New York: New York University Press.
Just, R. E., D. L. Hueth, and A. Schmitz. 1982. *Applied Welfare Economics and Public Policy*. Englewood Cliffs: Prentice-Hall.
Kahneman, D. and A. Tversky. 1972. 'Subjective Probability: A Judgment of Representatives.' *Cognitive Psychology* 3(3): 430–454.
——— and ———. 1979. 'Prospect Theory: An Analysis of Decisions under Risk.' *Econometrica* 47(2): 263–291.
——— and ———. 1982. 'The Psychology of Preferences.' *Scientific American* January: 160–173.
———, P. Slovic, and A. Tversky. 1982. *Judgement under Uncertainty: Heuristics and Biases*. Cambridge: Cambridge University Press.
Kleindorfer, P. R. and H. Kunreuther. 1983. 'Misinformation and Equilibrium in Insurance Market.' In J. Finsinger, ed., *Issues in Pricing and Regulation*. New York: Lexington.
Kneese, A. V. 1962. 'Water Pollution: Economic Aspects in Research Needs.' Washington D.C.: Resources for the Future.
———. 1964. *The Economics of Regional Water Quality Management*. Baltimore: Johns Hopkins University Press.
Knetsch, J. L. and R. K. Davis. 1966. 'Comparisons of Methods for Recreation Evaluation.' In *Water Research*, A. V. Kneese and S. C. Smith, eds. Baltimore: Johns Hopkins University Press.
——— and J. A. Sinden. Forthcoming. 'Willingness to Pay and Compensation Demanded: Experimental Evidence of an Unexpected Disparity in Measures of Value.' *Quarterly Journal of Economics*.
Knight, F. H. 1951. *The Economic Organization*. New York: Augustus Kelley.
Kogan, N. and M. A. Wallach. 1964. *Risk Taking: A Study in Cognition and Personality*. New York: Holt, Rinehart and Winston.
Krutilla, J. V. 1967. 'Conservation Reconsidered.' *American Economic Review* 57(4): 777–786.
Kunreuther, H. 1976. 'Limited Knowledge and Insurance Protection.' *Public Policy* 24(2): 227–261.
———, R. Ginsberg, L. Miller, P. Sagi, P. Slovic, B. Borkan, and N. Katz. 1978. *Disaster Insurance Protection: Public Policy Lessons*. New York: Wiley.
Lancaster, K. J. 1966. 'A New Approach to Consumer Theory.' *Journal of Political Economy*: 132–157.
LaPiere, R. T. 1934. 'Attitudes vs. Actions.' *Social Forces* 13.
Leamer, E. E. 1983. 'Let's Take the Con Out of Econometrics.' *American Economic Review* 73: 31–43.
Lichtenstein, S. and P. Slovic. 1971. 'Reversal of Preferences between Bids and Choices in Gambling Decisions.' *Journal of Experimental Psychology* 89: 46–55.
——— and B. Fischhoff. 1978. 'Judged Frequency of Lethal Events.' *Journal of Experimental Psychology* 4(6): 551–578.

Lipsey, R. L. and K. Lancaster. 1956. 'The General Theory of Second Best.' *Review of Economic Studies* 24(1).

Little, I. M. D. 1952. 'Social Choice and Individual Values.' *Journal of Political Economy* 60(5).

Loeb, M. 1977. 'Alternative Versions of the Demand-Revealing Process.' *Public Choice* 29 (Special Supplement to Spring).

Loehman, E., S. Berg, A. Arroyo, R. Hedinger, J. Schwartz, M. Shaw, R. Fahien, V. De, R. Fishe, D. Rio, W. Rossley, and A. Green. 1979. 'Distributional Analysis of Regional Benefits and Cost of Air Quality Control.' *Journal of Environmental Economics and Management* 6: 222–243.

Loomis, J. B. and R. G. Walsh. 1982. 'An Economic Evaluation of the Benefits and Costs of U.S. Forest Service Wilderness Study Areas in Colorado.' Sec. II of *Colorado's Wilderness Opportunity*. Denver: American Wilderness Alliance.

Lucas, R. E. B. 1977. 'Hedonic Wage Equation and Psychic Wages in the Return to Schooling.' *American Economic Review* 67: 549–558.

Luce, R. D. 1956. 'Semiorder and a Theory of Utility Discrimination.' *Econometrica* 24.

Machlup, F. 1967. 'Theories of the Firm: Marginalist, Behavioral, Managerial.' *American Economic Review* 57(1).

Majid, J., J. A. Sinden, and A. Randall. 1983. 'Benefit Evaluation of Increments to Existing Systems of Public Facilities.' *Land Economics* 59(4): 377–392.

Maler, J. 1977. 'A Note on the Use of Property Values in Estimating Marginal Willingness to Pay for Environmental Quality.' *Journal of Environmental Economics Management* 4 (December): 355–369.

Maler, K. G. 1974. *Environmental Economics*. Baltimore: Johns Hopkins University Press.

McConnell, K. E. and N. E. Bockstael. 1984. 'Aggregation in Recreation Economics: Issues of Estimation and Benefit Measurement.' *Procedures of the Annual Meeting of the American Association of Agricultural Economics*. Cornell University.

McMillan, J. 1979. 'The Free-Rider Problem: A Survey.' *The Economic Record* (June): 95–107.

McNeil, B. J. 1982. 'On the Elicitation of Preferences for Alternative Therapies.' *New England Journal of Medicine* 306 (May): 1259–1262.

——, S. J. Panker, H. C. Cox, and A. Tversky. 1982. 'Patient Preferences for Alternative Therapies.' Unpublished manuscript.

Mendelsohn, R. 1980. 'The Demand and Supply for Characteristics of Goods.' Mimeographed manuscript. University of Washington.

—— and G. M. Brown, Jr. 1983. 'Revealed Preference Approach to Valuing Outdoor Recreation.' *Natural Resources Journal* 23(3): 607–618.

Milgrom, P. R. and R. J. Weber. 1982. 'A Theory of Auctions and Competitive Bidding.' *Econometrica* 50 (5): 1089–1122.

Miller, G. A. 1956. 'The Magical Number Seven, Plus or Minus Two; Some Limits on our Capacity for Processing Information.' *Psychology Review* 63(2): 81–97.

Miller, G. and C. Plott. 1983. 'Revenue Generating Properties of Sealed-Bid Auctions.' In V. L. Smith, ed., *Research in Experimental Economics*, Vol III. JAI Press

Mitchell, R. C. and R. T. Carson. 1981. *An Experiment in Determining Willingness to Pay for National Water Quality Improvements*. Draft report prepared for the Environmental Protection Agency. Washington, D. C.: Resources for the Future.

Morgenstern, O. 1973. *On the Accuracy of Economics Observations*. 2nd ed. Princeton: Princeton University Press.

Ness, H. O. 1963. *Market Potential for Selected Fee Hunting Enterprises in New Mexico*. Unpublished M. A. Thesis. New Mexico State University.

O'Hanlon, P. and J. Sinden. 1978. 'Scope for Valuation of Environmental Goods: Comment.' *Land Economics* 54(3): 381–387.

Palfrey, T. R. 1980. 'Equilibrium Models of Multiple-Object Auctions.' Unpublished Ph.D. dissertation. California Institute of Technology.

Plott, C. R. 1979. 'The Application of Laboratory Experimental Methods to Public Choice.' In C. S. Russell, ed., *Collective Decision Making.* Baltimore: Johns Hopkins University Press.

———. 1982. 'Industrial Organization Theory and Experimental Economics.' *Journal of Economic Literature* 20.

Pommerehne, W., S. Schneider, and P. Zweifel. 1982. 'Economic Theory of Choice and the Preference Reversal Phenomenon: A Reexamination.' *American Economic Review* 72 (June): 569–574.

Pratt. J. W., D. Wise and R. Zeckhauser. 1979. 'Price Differences in Almost Competitive Markets.' *Quarterly Journal of Economics* 93(2): 189–211.

Quigley, J. M. 1982. 'Nonlinear Budget Constraints and Consumer Demand: An Application of Public Programs for Residential Housing.) *Journal of Urban Economics* 12: 177–201.

Rae, D. A. 1983. 'The Value to Visitors of Improving Visibility at Mesa Verde and Great Smokey National Parks.' In R. D. Rowe and L. G. Chestnut, eds., *Managing Air Quality and Scenic Resources at National Parks and Wilderness Areas.* Boulder: Westview.

Randall, A. n.d. 'The Possibility of Satisfying Benefit Estimation with Contingent Markets.' Mimeographed manuscript. University of Kentucky.

———, O. Grunwald, S. Johnson, R. Ausness, and R. Pagoulatos. 1978. 'Reclaiming Coal Surface Mines in Central Appalachia: A Case Study of the Benefits and Costs.' *Land Economics* 54(4): 472–489.

———, ———, A. Pagoulatos, R. Ausness, and S. Johnson. 1978. *Estimating Environmental Damages from Surface Mining of Coal in Appalachia: A Case Study.* Report to the Environmental Protection Agency.

———, J. P. Hoehn, and D. S. Brookshire. 1983. 'Contingent Valuation Surveys for Evaluating Environmental Assets.' *Natural Resources Journal* 23(3): 635–648.

———, ———, and G. S. Tolley. Forthcoming. 'The Structure of Contingent Markets: Some Experimental Results.' *Journal of Environmental Economics and Management.*

———, B. Ives, and C. Eastman. 1974a. 'Benefits of Abating Aesthetic Environmental Damage from the Four Corners Power Plant, Fruitland, New Mexico.' *New Mexico State University Agricultural Experiment Station Bulletin* 618.

———, ———, ———. 1974b. 'Bidding Games for Valuation of Aesthetic Environmental Improvements.' *Journal of Bidding Economics and Management* 1: 132–149.

Reilly, R., 1982. 'Preference Reversal: Further Evidence and Some Suggested Modifications in Experimental Design.' *American Economic Review* 72 (June): 576–584.

Ridker, R. 1967. *Economic Costs of Air Pollution.* New York: Praeger.

Robbins, L. 1982. *An Essay on the Nature and Significance of Economic Science.* London: Macmillan.

Robertson, L. 1974. 'Urban Area Safety Belt Use in Automobiles with Starter Interlock Belt Systems: A Preliminary Report.' Washington, D. C.: Insurance Institute for Highway Safety.

Robinson, J. 1962. *Economics Philosophy.* Garden City: Doubleday.

Ronan, J. 1973. 'Effects of Some Probability Displays on Choices.' *Organizational Behavior* 9(1): 1–15.

Rosen, S. 1974. 'Hedonic Prices and Implicit Market: Product Differentiation in Pure Competition.' *Journal of Political Economy* 82: 34–55.

Rothenburg, J. 1961. *The Measurement of Social Welfare.* Englewood Cliffs: Prentice Hall.

Rowe, R., R. C. d'Arge, and D. S. Brookshire. 1980a. 'Environmental Preferences and Effluent Charges.' In *Progress in Resource Management and Environmental Planning,* Vol. II, T. O'Riordan and K. Turner, eds. New York: Wiley.

———, ———, ———. 1980b. 'An Experiment on the Economic Value of Visibility.' *Journal of Environmental Economics and Management* 7: 1–19.

——— and L. G. Chestnut. 1983. 'Valuing Environmental Commodities: Revisited.' *Land Economics* 59(4): 404–410.

Samuelson, P. A. 1954. 'Pure Theory of Public Expenditure.' *Review of Economics and Statistics* 36 (November): 387–389.
———. 1955. 'Diagrammatic Exposition of a Theory of Public Expenditure.' *Review of Economics and Statistics* 37.
———. 1958. 'Aspects of Public Expenditure Theories.' *Review of Economics and Statistics.* 37.
Scheffe, H. 1970. 'Practical Solutions to the Behrens-Fisher Problem.' *Journal of the American Statistical Association* 65 (December).
Schelling, T. C. 1978. *Micromotives and Macromotives.* New York: Norton.
Scherr, B. A. and E. M. Babb. 1975. 'Pricing Public Goods: An Experiment with Two Proposed Pricing Systems.' *Public Choice* 23: 35–48.
Schoemaker, P. J. H. 1980. *Experiments on Decisions under Risk: The Expected Utility Hypothesis.* Boston: Nijhoff.
———. 1982. 'The Expected Utility Model: Its Variants, Purposes, Evidence and Limitations.' *Journal of Economic Literature* 20 (June): 529–563.
——— and H. C. Kunreuther. 1979. 'An Experimental Study of Insurance Decisions.' *Journal of Risk and Insurance* 46(4): 603–618.
Schulze, W. D., R. C. d'Arge, and D. S. Brookshire. 1981. 'Valuing Environmental Commodities: Some Recent Experiments.' *Land Economics* 57: 151–172.
———, D. S. Brookshire, and T. Sandler. 1981. 'The Social Rate of Discount for Nuclear Waste Storage: Economics or Ethics?' *Natural Resources Journal* 21(4): 811–832.
———, ———, E. G. Walther, K. Kelley, M. A. Thayer, R. L. Whitworth, S. Ben-David, W. Malm, and J. Molenar. 1981. 'The Benefits of Preserving Visibility in the National Parklands of the Southwest.' Vol. III. *Methods Development for Environmental Control Benefits Assessment.* Final report to the Environmental Protection Agency.
———, ———, ———, K. K. MacFarland, M. A. Thayer, R. L. Whitworth, S. Ben-David, W. Malm, and J. Molenar. 1983. 'The Economic Benefits of Preserving Visibility in the National Parklands of the Southwest.' *Natural Resources Journal* 23: 149–173.
———, R. G. Cummings, D. S. Brookshire, M. A. Thayer, R. Whitworth, and M. Rahmatian. 1983. 'Experimental Approaches to Valuing Environmental Commodities: Vol. II.' Draft final report for *Methods Development in Measuring Benefits of Environmental Improvements* for the Environmental Protection Agency.
Schuman, H. and M. P. Johnson. 'Attitudes and Behavior.' *Annual Review of Sociology* 2 (1976): 161–207.
Seller, C., J. R. Stoll, and J.-P. Chaves. n.d. 'Validation of Empirical Measures of Welfare Change: A Comparison of Nonmarket Techniques.' Natural Resources Working Paper Series. Department of Agricultural Economics, Texas A & M University.
Shubik, M. 1975. 'Oligopoly Theory, Communication and Information.' *American Economic Review* 65.
Simon, H. A. 1955. 'A Behavioral Model of Rational Choice.' *Quarterly Journal of Economics* 69(1): 174–183.
———. 1979. 'Rational Decision Making in Business Organizations.' *American Economic Review* 69(4): 493–513.
——— and A. Newell. 1971. 'Human Problem Solving: The State of the Theory in 1970.' *American Psychology* 26(2): 145–159.
Sinden, J. and J. Wyckoff. 1976. 'Indifference Mapping: An Empirical Methodology for Economic Evaluation of the Environment.' *Regional Science and Urban Economics* 6 (March): 81–103.
Slovic, P. 1969. 'Differential Effects of Real Versus Hypothetical Payoffs on Choices among Gambles.' *Journal of Experimental Psychology* 80(3): 434–437.
———, B. Fischhoff and S. Lichtenstein. 1980. 'Facts and Fears: Understanding Perceived Risk.' In R. C. Schwing and W. A. Albers, Jr., eds, *Societal Risk Assessment: How Safe is Safe Enough?* New York: Plenum.

———, H. Kunreuther, and G. F. White. 1974. 'Decision Processes, Rationality and Adjustment to Natural Hazards.' In G. F. White, ed., *Natural Hazards.* Oxford: Oxford University Press.

——— and S. C. Lichtenstein. 1968. 'Relative Importance of Probabilities and Payoffs in Risk Taking.' *Journal of Experimental Psychology* 78 (3, Part 2).

——— and A. Tversky. 1974. 'Who Accepts Savages Axioms?' *Behavioral Science* 19(6): 368–373.

Smith V. K. 1984. 'To Keep or Toss the Contingent Valuation Method.' Mimeographed manuscript. Department of Economics, Vanderbilt University.

Smith, V. L. 1967. 'Experimental Studies of Discrimination vs. Competition in Sealed-Bid Auction Markets.' *Journal of Business* 40.

———. 1976. 'Experimental Economics: Induced Value Theory.' *American Economic Review Proceedings* 66.

———. 1977. 'The Principle of Unanimity and Voluntary Consent in Social Choice.' *Journal Political Economy* 85: 1125–1139.

———. 1979a. 'Incentive Compatible Experimental Processes for the Provision of Public Good.' In V. L. Smith, ed., *Research in Experimental Economics,* Vol. 1, JAI Press.

———. 1979b. 'An Experimental Comparison of Three Public Good Decision Mechanisms. *Scandinavian Journal of Economics* 81.

———. 1980. 'Experiments with a Decentralized Mechanism for Public Good Decisions.' *American Economic Review* 70: 584–599.

———. 1982. 'Microeconomic Systems as an Experimental Science.' *American Economic Review* 72(5): 923–955.

Stigler, G. 1950. 'The Development of Utility Theory: II.' *Journal of Political Economy* 63(5): 373–396.

——— and G. S. Becker. 1977. 'De Gustibus non est Disputandum.' *American Economic Review* 67(2): 76–90.

Strand, J. n.d. 'Economic Evaluation of Damage to Freshwater Fish in Norway due to Acid Precipitation.' Institute of Economics, University of Oslo, Norway.

Strauss, R. P. and G. D. Hughes. 1976. 'A New Approach to the Demand for Public Goods.' *Journal of Public Economics* 6 (October): 191–204.

Sutherland, R. J. and R. G. Walsh. 1982. 'Effect of Distance on the Preservation Value of Water Quality.' Journal draft.

Thaler, R. 1980. 'Toward a Positive Theory of Consumer Choice.' *Journal of Economic Behavior and Organization* 1 (March): 39–60.

——— and S. Rosen. 1975. 'The Value of Saving a Life: Evidence from the Labor Market.' In N. Terleckys, ed., *Household Production and Consumption.* New York: Columbia University Press.

Thayer, M. A. 1981. 'Contingent Valuation Techniques for Assessing Environmental Impacts: Further Evidence.' *Journal of Environmental Economics and Management* 8: 27–44.

Tideman, N., and G. Turlock. 1976. 'A New and Superior Process for Making Social Choices.' *Journal of Political Economy* 84.

Tolley, G. S., A. Randall, G. Blomquist, R. Fabian, G. Fishelson, A. Frankel, J. P. Hoehn, R. Krumm, and E. Mensah. Forthcoming. *Establishing and Valuing the Effects of Improved Visibility in Eastern United States.* Final report for Environmental Protection Agency.

Tversky, A. and D. Kahneman. 1973. 'Availability: A Heuristic for Judging Frequency and Probability.' *Cognitive Psychology* 5(2): 207–232.

———. ———. 1974. 'Judgement under Uncertainty: Heuristics and Biases.' *Science* 185(2): 1124–1131.

———, ———. 1981. 'The Framing of Decisions and the Psychology of Choice.' *Science* 211: 453–458.

Ullman, E. and D. Volk. 1961. 'An Operational Model for Predicting Reservoir Attendance and Benefits.' *Proceedings of the Michigan Academy of Sciences.*

Vaughn, W. and C. Russell. 1982. *The National Benefits of Water Pollution Control: Fresh Water Recreational Fishing*. Baltimore: Johns Hopkins University Press.

Vickery, W. 1961. 'Counterspeculation, Auctions and Competitive Sealed Tenders.' *Journal of Finance* 16 (March): 8–37.

———. 1976. 'Auctions, Markets, and Optimal Allocation.' In Y. Amihud, ed., *Bidding and Auctioning for Procurement and Allocation*. New York: New York University Press.

Walsh, R. G. 1980. *An Economic Evaluation of the General Management Plan for Yosemite National Park*. Technical Report No. 19. Water Resources Research Institute, Colorado State University, Fort Collins.

———, R. Aukerman, and R. Milton. 1980. *Measuring Benefits and the Economic Value of Water in Recreation on High Country Reservoirs*. Completion Report No. 102. Water Resources Research Institute. Colarado State University, Fort Collins.

———, R. K. Ericson, and D. J. Arosteguy, with M. P. Hansen. 1980. *An Application of a Model for Estimating the Recreation Value of Instream Flow*. Completion Report No. 101. Water Resources Research Institute, Colorado State University, Fort Collins.

———, ———, J. R. McKean, and R. A. Young. 1978. *Recreation Benefits of Water Quality: Rocky Mountain National Park, South Platte River Basin, Colorado*. Technical Report No. 12. Environmental Resources Center, Colorado State University, Fort Collins.

———, D. A. Greenley, R. A. Young, J. R. McKean, and A. A. Prato. 1978. *Option Values, Preservation Values and Recreational Benefits of Improved Water Quality: A Case Study of the South Platte River Basin, Colorado*. Socioeconomic Environmental Studies Series. Environmental Protection Agency. Research Triangle Park, North Carolina.

———, G. Keleta, and J. P. Olienyk. 1981. *Value of Trees to Residential Property Owners with Mountain Pine Beetle and Spruce Budworn Damage in the Colorado Front Range*. Report by the Department of Economics to the Forest Service, Department of Agriculture.

——— and J. P. Olienyk. 1981. *Recreation Demand Effects of Mountain Pine Beetle Damage to the Quality of Forest Recreation Resources in the Colorado Front Range*. Report by the Department of Economics to the Forest Service, Department of Agriculture.

Walbert, M. S. Forthcoming. 'Valuing Policies which Reduce Environmental Risk: An Assessment of the Contingent Valuation Method.' Ph.D. dissertation. Department of Economics, University of New Mexico.

Ward, E. 1954. 'The Theory of Decision Making.' *Psychological Bulletin* 51(4): 380–417.

Weinstein, M. C. and R. J. Quinn. 1983. 'Psychological Considerations in Valuing Health Risk Reductions.' *Natural Resources Journal* 23(3): 659–673.

Weisbrod, B. A. 1964. 'Collective-Consumption Services of Individual-Consumption Goods.' *Quarterly Journal of Economics* 78 (August): 471–477.

Wicksell, K. 1896. 'A New Principal of Just Taxation.' Translated by J. M. Buchanan, in R. A. Musgrave and A. T. Peacock, *Classics in the Theory of Public Finance*. New York: St Martin's Press. 1967.

Wilde, L. 1980. 'On the Use of Laboratory Experiments in Economics.' In J. Pitt, ed., *The Philosophy of Economics*. Dordrecht: Reidel.

Willig, R. D. 1976. 'Consumer's Surplus without Apology.' *American Economic Review* 66(5): 89–97.

Chapter 7

The Valuation of Risks to Life and Health: Guidelines for Policy Analysis

W. Kip Viscusi

The task of valuing the impacts of government policies on life and health remains a particularly controversial undertaking. In part, this controversy has been stimulated by the special status accorded to health-related concerns, as reflected in the wide range of health-enhancing government programs. Unfortunately, there has also been much needless controversy stimulated by misconceptions regarding what the appropriate economic measure of the value of health status should be.

Initial efforts to value health focused on the direct monetary costs involved – the lost income, medical expenses, and other out-of-pocket outlays. Until recently, this approach was the dominant benefit assessment technique used by Federal agencies. The source of the popularity of the direct cost approach is that the task of benefit assessment becomes a well-defined problem. Data on worker wages are widely available, and a few minor assumptions enable one to obtain estimates of the economic costs of the health impacts. This precision is largely illusory to the extent that the analyst is not measuring the benefits of health risk reduction but rather a highly imperfect proxy for these broader concerns. In this review, I will outline a more appropriate procedure for such benefit assessments and the range of values associated with this approach.

Willingness-to-Pay Principles

The conceptual basis for valuing reductions in risks to life and health is the same as for other benefit categories. It is society's willingness to pay for this risk reduction that is the appropriate measure. Much of the benefit will be derived by those directly affected by the policy and by their families, but the altruistic concerns of society at large also enter. Whether or nor those bearing the risk are cognizant of the risk or are compensated for incurring it (e.g., through wage premiums for risky jobs) does not affect the rationale for using the willingness-to-pay approach, although it may affect the desirability of the policy and how much society is willing to pay to reduce the risk. In particular, the perceived equity of the risk may be greatly affected by the presence of compensation.

Ever since the classic essay by Schelling (1968) economists have focused, not on how much society should pay to prevent certain death, but on how much we are willing to pay to save statistical lives. In particular, the pertinent benefit of a government program is a reduction of the probability of death or some other health aspect for a large number of individuals rather than the prevention of a certain number of deaths that might be identified after the fact.

One reason for this distinction is that society's valuation of an identified certain life is likely to be greater than that of a statistical life. A trapped coal miner whose plight is featured on the evening news may evoke more public concern than the unidentifiable beneficiary of automobile passive-restraint systems. This difference in attitudes is usually taken as an index of society's excessive concern for identified lives. However, one might see the pattern somewhat differently if one viewed the high valuation of identified lives as being the result of a more thoughtful expression of individual values than in the airbag case, where the implications for risk reduction may be less clear-cut to the casual observer. One also might view it as reflecting a greater sense of responsibility for identifiable fatalities. The consensus in the literature is that society places an irrationally large weight on identified lives, but the interpretation of this phenomenon is by no means straightforward.

A second reason for drawing a distinction between statistical lives and certain lives can be traced to the role of wealth effects. An individual's willingness to pay to reduce a risk should be greater per unit risk for small risks than for large risks. A similiar principle determines one's attitude toward valuing large risks as opposed to small risks. If one were to purchase such a risk reduction sequentially by, for example, buying back bullets from a gun with which one is being forced to play Russian roulette, one would be willing to pay less for each successive bullet as each bullet was purchased because one becomes poorer with each bullet purchased.[1] For much the same reason, if a worker were willing to pay $2 to reduce the risk of his job by one chance in a million per lifetime, this result does not mean he would be willing to pay $2 million to prevent certain death. Such an allocation might greatly exceed his lifetime wealth.

To clarify the underpinnings of the value of risk reduction methodology, it may be instructive to summarize a model based on the job risk case considered in Viscusi (1979). In this instance, the value of life will be implied by the risk-dollars tradeoff selected by the worker.

Let p be the probability of an event that leads to one's death and $w(p)$ represent the schedule of annual earnings for jobs posing a risk p. The principal matter of interest is the slope of this relationship – the risk-dollars tradeoff. The increase in earnings in response to an increase in risk will be denoted by dw/dp (i.e., the derivative of earnings with respect to the risk probability). For the case of certain death, this rate of tradeoff represents the implicit value of one's life; this magnitude in turn hinges on worker attitudes toward risk. More specifically, let Y^0 represent initial assets, x represent consumption (equal to $Y^0 + w(p)$), U^1 represent utility when healthy, and U^2 represent utility when injured or dead, where $U^1(x) > U^2(x)$; $U^1_x > U^2_x$ and $U^1_{xx}, U^2_{xx} \leq 0$. Then one can show that

$$\frac{dw}{dp} = \frac{U^1 - U^2}{(1-p) U^1_x + p U^2_x} = \frac{\text{difference in welfare when healthy or injured}}{\text{expected marginal utility of compensation}}$$

If p represents the risk of death, dw/dp represents the implicit value per unit risk to one's life; for situations in which p represents the nonfatal risk, dw/dp represents the implicit value per unit risk of injury.

This equation makes clear the fact that the implicit value terminology is somewhat misleading, since dw/dp does not represent the amount the worker would require to accept certain death or injury. Rather, it reflects the worker's rate of tradeoff between risk and dollars for very small risks. According to this model, for sufficiently small risk changes, the risk-dollars tradeoff should be the same whether the individual is being paid to incur a greater risk or is spending money to purchase a reduction in risk. A worker who values his life at $1 million will require $100 to accept a 1 in 10,000 chance of death. Alternatively, if there were a group of 10,000 individuals, one of whom would be killed, then overall these individuals would accept one certain but randomly inflicted death if they were compensated $100 each.

This development has been based on the assumption that individuals are rational and, more specifically, on the assumption that they maximize subjective expected utility (SEU). Although convenient analytically, the basic result regarding willingness to pay for marginal reductions in risk does not require that the expected utility hypothesis hold. As shown by Arrow (1983), one obtains similar results if individuals have utility functions that depend on consumption (or money) and the risk level, where this utility function need only be defined up to a monotone transformation.

The resolution of lotteries over time may create anxiety for the individual incurring the risk. This factor can be included as an additional component in the formulation of the willingness-to-pay analysis,[2] but its practical import is unclear. The possibility of cognitive dissonance with respect to risk-induced anxiety also arises if people find it beneficial to ignore the risks they face (see Akerlof and Dickens 1981). If this selective attention to one's environment occurs after one has made a risk-related choice, then the earlier analysis is unchanged. If individuals do not make sensible choices with respect to risks because of cognitive dissonance, misperception of the risk, or some other type of irrationality, then the appropriate policy issue is how individuals would value the risk reduction *if* they were rational (see Chapter 5).

Adjustment in the Value of Life: Quantity, Quality, and Heterogeneity

The value attached to reducing risks to life and health will not be a single number. In the case of policies affecting individual's lives, these efforts do not confer immortality but simply extend an individual's life by preventing an immediate death. Saving the life of a 20-year-old will save a greater number of discounted expected life years than saving the life of an 80-year-old. At high discount rates such as the 10 percent rates now employed, these quantity adjustments will make little difference unless the age differences are stark. The discounted number of life years saved is about the same in each case.

The timing of the life extension is of consequence, however, in that one should discount deferred health benefits (in monetary terms) just as one would discount other deferred policy impacts. A frequently voiced counterargument is that a year of hearing loss at age 25, for example, is no less important than a year of hearing loss at age 20. Therefore, these health impacts should receive equal weight, even though one of these incidents occurs five years later. At the time of their occurrence, these health effects may be associated with equal willingness to pay (WTP). The willingness-to-pay amount for the more immediate loss should receive a greater weight, since this money could be invested, giving it a terminal value five years hence that is greater than the original willingness-to-pay figure. The practice of discounting the monetary value of deferred health benefits accomplishes a similar result.

The choice of the appropriate level of the discount rate is by no means clear-cut. Consumers' inability to borrow and lend at the same rate is both a cause and a reflection of the existence of capital market imperfections and implies that the discount rate may be different for different consumers. Insurance market imperfections also influence observed interest rates since individuals will save more to provide for future contingencies in the absence of perfect insurance markets. These issues have been the focus of debates in the benefit-cost analysis literature for decades, and the administrative solution has been for the Office of Management and Budget to prescribe a 10 percent discount rate.

The quality of the life extension also matters, and ideally one should focus on quality-adjusted life years (see Zeckhauser and Shepard 1976). If the lives saved involve reduced capacity (e.g., a stroke victim), one should take this lower life quality into account, since the individual's own valuation of such a potential life extension would be less under these circumstances.

These quality concerns are particularly pertinent with respect to life-extending efforts for individuals with catastrophic illnesses. In the case of shifts in mortality distributions for individuals in good health, the quantity adjustments and particularly the discounting of deferred benefits are most essential.

These concerns are within individual variations in how health impacts should be valued. Additional concerns arise across individuals because of heterogeneity in the value of life. There is no theoretical restriction that individuals should have the same value of life. In general, individuals would be expected to have quite different values of life that hinge on their attitudes toward risk, just as their other tastes vary.

One determinant of the monetary value of safety that people will have is their individual wealth. There is a strong negative relationship between individual wealth and the risks one will choose to accept. The richer one is, the safer the job one will select from any given wage-risk schedule that is offered in the market, other things being equal. Society may wish to redistribute income or undertake educational programs to boost earnings opportunities so that people will not find it necessary to increase their income through hazardous work, but in terms of the market choices these different tradeoffs reflect the preferences individuals exhibit.

There is no natural force that would drive individuals to have the same risk-money tradeoff. Whereas most consumer items command a single price, jobs are likely to command quite different risk premiums per unit of risk for different people. This is because the risk of a job is indivisible and cannot be spread across the entire population.

Workers who are informed of these risks before they choose their jobs should select the occupations that are most appropriate for their own preferences. Individuals with low values of safety consequently should accept high-risk jobs and receive lower premiums per unit of risk, whereas individuals putting high values on safety should accept low-risk jobs and receive higher risk premiums per unit of risk. In many studies it has not been possible to estimate this heterogeneity in the value of safety (i.e., in the value of statistical lives) because of the lack of sufficient empirical information. Nonetheless, this inability does not imply that this heterogeneity is unimportant.

In circumstances in which the heterogeneity in valuations of health can be identified, incorporating these differences in a policy analysis may be controversial. Some might question that reducing the fatality risks for cigarette smoking should be accorded a low value, but a low-income worker who attempts to boost his income through a job as risky as smoking cigarettes might be viewed somewhat differently. The reason for incurring the voluntary risk may be of consequence insofar as it may affect society's altruistic concerns.

If one were to use uniform values of health for all individuals in these situations rather than a lower value for those who are involved in largely voluntary, high-risk situations, in effect one would be valuing the high-risk individuals' lives by more than they would themselves. In the case of a regulation that alters the available job opportunities, the net effect may be to eliminate risk-dollar tradeoffs which the individuals themselves found attractive. To the extent that workers' decisions are rational, such regulations will lower their welfare. In the case of a program that provides medical services, a uniform value of life serves as an implicit means of redistribution. In doing so, it will be less efficient than a direct-transfer program in accomplishing this objective.[3]

The Human Capital Approach

Although the willingness-to-pay methodology is the generally accepted approach among economists, it is by no means the most prevalently used method for valuing health risks. For the past two decades the most popular approach has been to examine the financial costs associated with the health impact, principally medical costs and the present value of lost earnings. This technique, which is illustrated in the analysis by Rice and Cooper (1967), has been termed the human capital approach because it parallels that taken in the human capital literature developed by Theodore Schultz, Gary Becker, and others.

That branch of academic literature which deals with human resources does not attempt to value health but rather the economic implications of training and education. The assumption that monetary implications dominate the individual's choice of a job-training program is reasonable, but it is far less realistic to assume that foregone earnings are a good proxy for the value of a lost limb or one's life. Indeed, it is noteworthy that no leading economist associated with the academic research on human capital has ever espoused this approach to valuing health risks.

To the extent that the human capital approach has any merit, it is that it provides a technique for estimating willingness to pay. It is not a competing conceptual approach but rather an empirical approach to the value-of-life problem. The reason for its popularity among analysts is clear: it makes the analyst's task of valuing life a well-defined problem. Once equipped with a set of mortality tables and individual earnings, it is a straightforward task to calculate the human capital benefits measure. Unfortunately, the analyst is assessing an incorrect measure of benefits precisely, so that the results are not particularly meaningful. In terms of the nature of the results, the human capital approach leads to value-of-life estimates about an order of magnitude smaller than those that address the risk-dollars tradeoff directly.[4] Clearly, individual well-being goes far beyond its financial implications.

Valuations Using Market Data

Conceptual Background

In most areas of economic inquiry, the usual approach to obtaining empirical estimates is to gather market data to estimate the relationship of interest. In this case, there is no explicit price for health outcomes, since individuals do not explicitly purchase risk reductions in the marketplace. Health effects are, however, among the attributes of products and jobs that individuals select. Ideally one would like to isolate the risk-dollar tradeoffs that are implicit in these choices. Indeed, since the time of Adam Smith (1776), economists have observed that hazardous jobs will command compensating differentials.

The techniques for assessing the risk-dollar tradeoffs were originally devised to analyze *hedonic*, or quality-adjusted, prices for automobiles (Griliches 1971). The quality component that will be the focus here is the risk associated with the job or product. Most of the emphasis to date has been on the labor market, since the extensive data available on characteristics of jobs and workers make it feasible to disentangle premiums for job risks from premiums for other attributes of the job.

The standard approach is to specify an earnings equation and to identify the risk premiums, controlling for other factors that influence one's income. More specifically, let w be the workers' annual earnings, x_i ($i = 1, 2, \ldots, n$) be a series of explanatory variables (e.g., education and union status), p be the annual death risk, and q be the annual nonfatal injury risk. The general linear form of the earnings equations is:

$$w = \alpha + \sum_{i=1}^{m} \beta_i x_i + \gamma_0 p + \gamma_1 q + u$$

where u is a random error term. In some analyses, the dependent variable is the natural logarithm of w. The coefficient γ_0 of p represents what has been termed the implicit value of life (i.e., $\partial w/\partial p$), and γ_1 represents the implicit value of a nonfatal injury. In fact, these values represent the risk-dollar tradeoff, not the value the person would place on certain death or injury. To the extent that the job risk variables measure the worker's true risk with some random measurement error, the resulting value of life estimates will be biased downward.

A model of this type would include the following explanatory variables. The x_i capture income-related personal characteristics (i.e., age, race, sex, marital status, education, job experience) and job-related characteristics (i.e., unionization, industry, occupation, supervisory status, physical conditions, work speed). The inclusion of extensive nonpecuniary job characteristics variables is essential to ensure that the estimated values of γ_0 and γ_1 reflect premiums for risk rather than rewards for other, possibly correlated, unpleasant job attributes. Studies that have included such lists are those of Viscusi (1979), which is based on the workers' particular job, and Brown (1980), who linked the worker's occupation to a set of average characteristics. Indeed, most studies do not even include the nonfatal injury rate because of an inability to successfully disentangle the premiums for fatal and nonfatal risks.

Empirical Estimates

The principal studies of market premiums for risk are summarized in Table 7–1. With the exception of Blomquist's (1979) analysis of seatbelt usage and Portney's (1981) study of the effect of air pollution on property values, the analyses all use labor market data. Those nonlabor market studies are more indirect. Blomquist, for example, had to construct an elaborate model to link the seatbelt buckling decision and its discomfort to the value of life. Portney's study utilized air pollution data, dose-response information, and property value data to infer a price-risk tradeoff. This analysis assumed that individuals are aware of these linkages and that it is the mortality risk that is instrumental. The labor market studies utilize an industry-specific risk level which should be a more direct index of the risks.

The usual procedure is to match workers' jobs or occupational classification to some measure of the risk using occupational fatality data based on life insurance or Bureau of Labor Statistics industry rate measures. Occupational risk measures have the advantage that the risks of occupations may be more similar across industries than are risks of a job within an industry. However, there has been no empirical verification that this is the case. In addition, life insurance data reflect non-job-related risks of death as well as those associated with the job, which is a disadvantage.

Ideally, one would like to use workers' subjective assessments of the risk rather than an objective proxy for the risk level, since it is the individual's own risk beliefs that will drive market outcomes. The University of Michigan Survey of Working Conditions data (see Viscusi 1979) utilize a danger-perception dummy variable that takes on a value of one if the worker believes that his job exposes him to dangerous or unhealthy conditions, and zero if it does not. Although this variable provides only a rough approximation to the actual risk level, the derived annual risk premium was almost identical whether one used the subjective risk perception data or the Bureau of Labor Statistics industry injury risk.

Of the labor market studies, the principal outlier is by Thaler and Rosen (1976), whose value of life estimate of $580,000 is the lowest from the labor market. This result is not spurious but is attributable primarily to the nature of their sample. The Thaler-Rosen results pertain to workers in very high risk jobs with an average annual risk of fatality of 1/1,000. In contrast, the average job poses an annual fatality risk of 1/10,000. Because of the

7. The Valuation of Risks to Life and Health

Table 7-1. Summary of market studies of risk tradeoffs*

Investigator	Sample	Implicit value of life	Implicit value of nonfatal injuries
Blomquist (1979)	Seatbelt usage, Panel Study of Income Dynamics, 1972	$560,000	–
Brown (1980)	National Longitudinal Survey, 1967–1973	$1–$1.5 million	–
Leigh (forthcoming)	Panel Study of Income Dynamics, 1974	$3.8–$8.9 million*	$45,000–$56,000
	Quality of Employment Survey, 1977	$4.8–$8.4 million*	$38,000–$64.000
Olson (1981)	Current Population Survey, 1973	$7.4 million	–
Portney (1981)	Air Pollution and Property Values	$593,000–$890,000	–
Smith (1965)	Current Population Survey, 1967	$7.5 million	–
	Current Population Survey, 1973	$3.3 million	–
Thaler and Rosen (1976)	Survey of Economic Opportunity, 1967	$580,000	–
Viscusi (1979)	Survey of Working Conditions, 1970–71	$2.9–$3.9 million	$23,000–$34,000
Viscusi (1981)	Panel Study of Income Dynamics, 1976	$7–$11 million*	$32,000–$35,000
Viscusi and O'Connor (1984)	Survey of Chemical Industry Workers	–	$10,000–$13,000

*All prices are in 1982 dollars. The asterisked results for the Leigh and Viscusi studies are evaluated at the mean risk level for the sample for models in which the heterogeneity in wage-risk tradeoffs was assessed. The appropriate implicit values were calculated from the regression equations using the appropriate $\partial w/\partial p$ and $\partial w/\partial q$ values discussed in the text.

heterogeneity of values of individual risk and self-selection of workers with low values into high-risk jobs, one should expect to obtain a lower value of life using a sample of workers in risky jobs.

For purposes of policy analysis, the appropriate value of life depends on whose values we wisk to assess. If a regulation affects individuals who have voluntarily put themselves into high-risk activities, a value of life around $600,000 seems appropriate. For risks that are more modest or have been incurred involuntarily, the appropriate value of life will be that of a more representative worker, which is in the range of about $2 to $3 million or more. Very small, involuntary risks, such as those associated with nuclear or airline safety, may command much higher values.[5]

The only other health outcomes that have been assessed using market wage studies are nonfatal work accidents. These accidents involved one or more lost workdays in about half the cases. Once again, there is a reasonably broad range of estimates.[6] The findings by Leigh (forthcoming) may be somewhat higher than the previous studies suggested, since he did not include a detailed set of nonpecuniary job characteristic variables. Therefore, his injury risk variable will reflect premiums for omitted job attributes correlated with riskiness. An estimate of a value of a nonfatal injury of about $20,000 to $30,000 might best be viewed as the consensus injury value range based on these studies.

The market risk studies have at least indicated the general order of magnitude of individuals' monetary valuations of life and injury. Perhaps the greatest research need is to expand the scope of health effects for which benefit values can be determined.

Nonmarket Valuations

Court Awards and Workers' Compensation

Work and seatbelt decisions are not the only choices involving risks from which one might attempt to impute a value of life. This literature began with an examination of life insurance policies (see Eisner and Strotz 1961). Life insurance does not, however, affect individual health. Instead it is directed at providing compensation after one's death. Examination of life insurance policies may provide information regarding the optimal bequest, but it does not indicate how much it is worth to extend lives.

Settlements in court cases tend to have a similar focus since the emphasis has been on compensating victims for monetary losses, principally foregone wages and medical expenses, rather than for pain and suffering. These nonmonetary aspects may be the dominant concern in the case of severe health impacts. The average bodily injury payment for fatalities in product liability cases is about $212,000 (1982 prices), which is about one-tenth or less than the values of life obtained in studies of market risk premiums.[7]

Nevertheless, there are some instances, particularly those involving scarring and brain damage, in which court awards may provide some guidance regarding the implicit value attached to nonmonetary aspects of health outcomes.[8] Table 7-2 summarizes the distribution of bodily injury payments for various types of health outcomes. What is most noteworthy is the comparatively small magnitudes involved, except in cases such as brain damage, quadriplegia, and paraplegia where the medical expenses are particularly high. These product liability payments reflect not only jury verdicts, but also out-of-court settlements, which generally are for consider-

7. The Valuation of Risks to Life and Health

Table 7–2. Distribution of payments by injury diagnosis, bodily injury cases

Injury diagnosis	Percent of successful claimants with this injury	Average payment
Amputation	2.6%	$112,988
Asphyxiation	1.0	69,787
Bruise-abrasion	3.8	5,165
Burn	7.6	78,786
Concussion	0.7	32,479
Dermatitis	2.1	1,468
Dislocation	0.3	32,120
Electrical shock	0.3	31,728
Fracture	16.7	21,146
Laceration	14.5	11,240
Poisoning	16.1	1,102
Strain–sprain	3.4	25,198
Disease–respiratory	0.6	59,621
Disease–cancer (including Hodgkins disease, leukemia)	0.3	166,883
Disease–other	0.9	17,414
Paraplegia	0.1	319,620
Quadriplegia	0.1	505,355
Brain damage	0.8	357,482
Other	28.1	16,127
Total	100.0%	$25,680
Unknown	–	$39,592

Source: Insurance Services Office (1977: 116). All figures are in 1977 prices.

ably lower amounts. Exploration of the jury verdict subsample may be instructive in providing insight into the values society at large attaches to various health outcomes.

Workers' compensation benefit levels also provide a potentially useful source of information on the valuation of various health outcomes (Oi 1973). As the payment schedules in Table 7-3 indicate, there appears to be very little attempt to fully reflect the values from the standpoint of prevention, which include the health impact as well as the monetary consequences. The loss of a thumb, for example, has an upper benefit limit not unlike the average value of a work accident involving at least one lost workday, which will typically be much less severe than losing one's thumb. This disparity is to be expected. The appropriate value of life and health from the standpoint of compensation will generally be less than for prevention. As a result, analysis of levels of *ex post* compensation will provide only an underassessment of the value of prevention.

A final possible source of information for assessing the appropriate value of health risk prevention is to examine the values implied by past societal decisions.[9] Although potentially instructive as a very rough means for

Table 7-3. Income benefit range under workers' compensation payments, 1983

Scheduled injuries	Low	High
Arm at shoulder	$10,000	$125,460
Hand	8,675	102,510
Thumb	3,250	31,248
First finger	1,800	18,257
Second finger	1,350	15,624
Third finger	924	12,475
Fourth finger	600	11,475
Leg at hip	9,360	125,460
Foot	6,000	81,340
Great toe	1,200	19,960
Other toes	480	11,475
Eye	6,000	84,150
Hearing (one ear)	2,000	24,950
Hearing (both ears)	8,000	87,325

Source: U.S. Chamber of Commerce (1983).

assessing the opportunity cost in terms of the performance of a program compared with other life-enhancing efforts, this technique is based on the implausible assumption that past decisions were optimal. Most legislative mandates of risk regulation agencies are not based on a balancing of the costs and benefits to ensure that policies maximize society's welfare; they are much narrower in scope. Therefore, it is inappropiate to analyze past decisions as if such an optimization were taking place.

There are also numerous practical difficulties in terms of comparability. Since agencies have used different discount rates and underlying technical assumptions, the results often are not comparable across different policies.[10] In addition, risk-reducing policies typically have multiple health impacts, not simply one. Since no previous analysis has attempted to make the necessary adjustments to take these multiple effects into account, this produces a possibly spurious variation in the apparent efficacy of different agencies' programs.

Interviews and Contingent Valuations

Risk-dollar tradeoffs based on actual decisions offer the advantage that these choices presumably reflect individuals' underlying preferences. An alternative, more direct approach is simply to ask individuals what their tradeoffs are or how much they would be willing to pay for a particular risk reduction (Jones-Lee 1976). This interview approach has gained increasing popularity recently in the environmental area, where it has been dubbed contingent valuation and is now being applied to health risks by several research groups.[11] Although some of these efforts are focusing on traditional concerns, such as mortality risks, others are addressing consumers' valuations of hand burns, choramine gas poisonings, and poisonings to children from bleach and liquid drain opener.

The first study of this type was by Acton (1973), who surveyed individuals' willingness to pay for improved ambulance service for heart attack victims. This study, which was innovative in many respects, produced a comparatively low value of life – under $100,000. This result may be an aberration due to the small sample size (36) or the focus on post-heart attack lives, which should be valued less. A more likely possibility is that individuals did not think carefully about the inherently difficult questions concerning willingness to pay for improved ambulance service and its link to their health.

Another potential limitation of interviews is that the subject may misrepresent his preferences either to impress the interviewer or to influence policies affected by the numbers. For familiar strategic reasons a respondent might have the incentive to misrepresent his willingness to pay for publicly supported risk reduction efforts, particularly if he will not be required to back up his responses financially. Appropriately structured questionnaires can reduce this difficulty.

Interview techniques have inherent shortcomings and may not be a good substitute for market risk premium analyses when these are feasible. However, if undertaken in a manner that closely resembles market behavior, interviews do represent a potentially promising mechanism for expanding the range of health outcomes for which we can assess the benefits.

An Example: OSHA Chemical Labeling

This section provides a brief illustration of a derivation of the value of life and health based on an assessment of the OSHA chemical labeling regulation.[12]

This illustration will not be a benefit-cost analysis but rather a cost-effectiveness analysis, since benefit-cost tests are frequently ruled out by agencies' legislative mandates. In addition, the appropriate value of the health outcome is often a key and controversial parameter. Rather than bury the health valuation assumption in a complex analysis, it seems preferable to indicate the cost per unit health impact, which policy makers can then compare with results such as those in Table 7-1 to ascertain whether the program is reasonable.

In the case of the chemical labeling standard there was an additional major uncertainty. OSHA claimed that the standard would eliminate 10 percent of all chemical-related injuries and illnesses, whereas OMB viewed a 5 percent figure as being more reasonable. Table 7–4 provides a sensitivity analysis using these two assumptions. The first line of the table lists the discounted costs of the regulation less all discounted benefits that have been monetized. By netting out all nonhealth impacts, such as the reduced

Table 7-4. Health impacts prevented and cost health impact

	Lost work-day equivalents			
	Weights – 1, 1, 20* Effectiveness		Weights – 1, 5, 20* Effectiveness	
	5%	10%	5%	10%
Net discounted costs less monetized benefits	2.632×10^9	2.616×10^9	2.632×10^9	2.616×10^9
Total lost work-day equivalents (discounted)	9.5×10^4	18.9×10^4	24.7×10^4	49.7×10^4
Net discounted cost/lost work-day equivalent	$27,900	$14,000	$10,700	$5,300

*These are the relative weights placed on lost work-day cases (always 1), disabling illnesses (1 or 5), and cancers (always 20) in constructing a measure of lost work-day equivalents.

costs of fires, the analysis can address the cost and health concerns alone.

Doing so is not straightforward, however, since there are three general classes of health effects – accidents, disabling illnesses, and cancer – to be placed on a common metric. Suppose, as a tentative value judgment, that a case of cancer has value comparable to the value of life of a worker in a high-risk job. In this case it should receive a benefit value roughly 20 times that of a lost work-day accident. As a result, preventing a discounted case of cancer would be treated as equivalent to preventing 20 work-day accidents. One could potentially justify a relatively higher cancer weight, but doing so will not be necessary for regulation to be desirable. In this case, the object was to perform a sensitivity test to see whether the regulation was desirable. Use of higher cancer values, which may increase benefits by a factor of 5 or more, will increase benefits accordingly.

The relative severity of disabling illnesses due to chemical illnesses is less clear, so weights of 1 and 5 lost work-day case equivalents were used, implying a value between $30,000 and $150,000. Under all assumptions outlined in Table 7-4, the cost-effectiveness is at or above the levels needed to issue the regulation, which the Reagan administration did.

The two principal ingredients in this analysis that may be useful for future studies are the use of a cost-effectiveness framework and a sensitivity analysis that attempts to weight multiple health impacts. The standard approach of calculating magnitudes, such as the cost per life, implicitly ignores all other health implications and may provide a misleading index of a policy's efficacy.

Conclusions

Despite the inherent sensitivity of valuing risks to life and health, these tasks are not beyond the capabilities of policy analysis. The underlying theoretical issues have been examined in detail, and there is a consensus among economists that the willingness-to-pay approach is appropriate for valuing health risks. In the case of some risks, particularly those for mortality, a good deal of research has already been done. A major task is to expand the scope of health impacts for which we have benefit values and to better utilize the findings we do have in policy analyses.

It is noteworthy how far the policy approach to these issues has advanced in recent years. For the past few years, agencies such as EPA and OSHA have introduced the willingness-to-pay measure into their regulatory analyses, basing these values on labor market estimates of the risk-dollar tradeoff.

Indeed, as of 1984, the valuation of life has become a generally accepted component of the debate over risk regulation. The recent debate over an OSHA construction industry standard epitomizes this change.[14] Rather than claiming that the value-of-life issue was too sensitive to be discussed, there was an open policy debate over the appropriate value of life. OSHA used a value of life of $3.5 million in its regulatory analysis based on results for the average blue-collar worker. OMB took a different approach, citing evidence regarding the heterogeneity in the value of life. After noting the high and well-known risks associated with construction jobs, OMB urged that OSHA use a lower value of life of $1 million. One Congressman viewed both of these estimates as too low, advocating a $7 million figure in line with results for the Panel Study of Income Dynamics. In each case, the willingness-to-pay approach was accepted, as was the importance of using labor market studies as a reference point.

Northwestern University

Notes

1. This Russian roulette analogy and its solution have been part of the oral tradition in the risk regulation area for over a decade. The originator was Zeckhauser (1975). See Weinstein, Shepard, and Pliskin (1980) for related work.

2. Although the most extensive analysis of anxiety effects and their relation to the marginal value of life is in Viscusi (1979), pivotal antecedents in the literature are the papers by Zeckhauser (1974) and Schelling (1968).

3. Hyland and Zeckhauser (1979) formalize this point for the general redistribution case.

4. Such a gap is to be expected. See Shepard and Zeckhauser (1984).

5. During a discussion at the Users Workshop, a consensus of the participants indicated a range of $3-$7 million.

6. Olson (1981) also finds nonfatal injury premiums but does not report an implicit value.

7. The payment data in 1977 prices appear in the Insurance Services Office (1979:113).

8. Scarring appears to be the chief health-related variable that has an impact on product liability settlements beyond its monetary implications.

9. See, e.g., Broder and Morrall (1983), Graham and Vaupel (1983), Viscusi (1983), Miller and Yandle (1979), and Morrall (forthcoming).

10. A principal exception is the careful analysis by Broder and Morrall (1981).

11. For a critique and review of this approach, see Cummings, Brookshire, and Schulze (1984).

12. W. Kip Viscusi, 'Analysis of OMB and OSHA Evaluations of the Hazard Communication Proposal.' Report Prepared for Secretary of Labor Donovan, March 15, 1982. The assumptions altered from those in the OSHA analysis included, among others: the effectiveness assumption, the incidence of disabilities, the fraction of cancer cases affected by labeling, medical cost inflation, and time lags before there is an effect on diseases that occur with a lag.

13. See the *New York Times,* March 23, 1984, p.13 (national edition) for discussion of the regulation and the role of this analysis in Vice President Bush's decision to issue the regulation despite OMB's earlier objections.

14. See the *Washington Post,* Oct. 24, 1984, p.A15, and the *New York Times,* Oct. 26, 1984, p.8 (national edition).

References

Acton, Jan. 1973. *Evaluating Programs to Save Lives: The Case of Heart Attacks.* Santa Monica: Rand Corporation.
Akerlof, George and William Dickens. 1981. 'The Economic Consequences of Cognitive Dissonance.' *American Economic Review* 72(3): 307–319.
Arrow, Kenneth J. 1983. 'Behavior under Uncertainty and Its Implications for Policy.' Technical Report No. 399. Center for Research on Organizational Efficiency. Stanford University.
Blomquist, Glenn. 1979. 'Value of Life Saving: Implications of Consumption Activity.' *Journal of Political Economy* 87: 540–558.
Broder, Ivy and John F. Morrall III. 1983. 'The Economic Basis for OSHA's and EPA's Generic Carcinogen Standard.' In *What Role for the Government?* R. Zeckhauser and D. Leebaert, eds. Durham: Duke University Press.
Brown, Charles. 1980. 'Equalizing Differences in the Labor Market.' *Quarterly Journal of Economics* 94(1): 113–154.
Cummings, R. G., D. S. Brookshire, and W. D. Schulze. 1984. 'Valuing Environmental Goods: A State of the Arts Assessment of the Contingent Valuation Method.' Draft report to the Environmental Protection Agency.
Eisner, Robert and Robert Strotz. 1961. 'Flight Insurance and the Theory of Choice.' *Journal of Political Economy* 69: 355–368.

Graham, John and James Vaupel. 1983. 'The Value of Life: What Difference Does It Make?' In *What Role for the Government?* R. Zeckhauser and D. Leebaert, eds. Durham: Duke University Press.

Griliches, Zvi, ed. 1971. *Price Indexes and Quality Change.* Cambridge, Mass.: Harvard University Press.

Hylland, Aanund and Richard Zeckhauser. 1979. 'Distributional Objectives Should Affect Taxes but Not Program Choice or Design.' *Scandinavian Journal of Economics* 264–284.

Insurance Services Office. 1979. *Product Liability Closed Claims Survey: A Technical Analysis of Survey Results,* pp. 264–284. New York: Insurance Services Office.

Jones-Lee, M. W. 1976. *The Value of Life: An Economic Analysis.* Chicago: University of Chicago Press.

Leigh, J. Paul. Forthcoming. 'Estimates of the Value of Accident Avoidance at the Job Depends on the Concavity of the Equalizing Differences Curve.' *Quarterly Review of Economics and Business.*

Miller, James C. III and Bruce Yandle, eds. 1979. *Benefit-Cost Analysis of Social Regulation.* Washington, D.C.: American Enterprise Institute.

Morrall, John F. III. Forthcoming. *OSHA after Ten Years.* Washington, D.C.: American Enterprise Institute.

Oi, Walter. 1973. 'An Essay on Workmen's Compensation and Industrial Safety.' In *Supplemental Studies for the National Commission on State Workmen's Compensation Laws.* Washington, D.C.: GPO.

Olson, Craig. 1981. 'An Analysis of Wage Differentials Received by Workers on Dangerous Jobs.' *Journal of Human Resources* 16(2): 167–185.

Portney, Paul, 1981. 'Housing Prices, Health Effects, and Valuing Reductions in Risk of Death.' *Journal of Environmental Economics and Management* 8: 72–78.

Rice, Dorothy and Barbara Cooper. 1967. 'The Economic Value of Life.' *American Journal of Public Health* 57: 1954–1966.

Schelling, Thomas. 1968. 'The Life You Save May be Your Own.' In *Problems in Public Expenditure Analysis,* S. Chase, ed. Washington, D.C.: Brookings Institution.

Shepard, Donald and Richard Zeckhauser. 1984. 'Survival Versus Consumption.' *Management Sciences* 30(4): 423–439.

Smith, Adam. 1776. *The Wealth of Nations.* Reprint ed., 1937. New York: Modern Library.

Smith, Robert S. 1976. *The Occupational Safety and Health Act: Its Goals and Achievements.* Washington, D.C.: American Enterprise Institute.

Thaler, Richard and Sherwin Rosen. 1976. 'The Value of Saving a Life: Evidence from the Labor Market.' In N. Terleckyj, ed., *Household Production and Consumption.* New York: Columbia University Press.

U.S. Chamber of Commerce. 1983. *Analysis of Workers' Compensation Laws.* Washington, D.C.: U.S. Chamber of Commerce.

Viscusi, W. Kip. 1979. *Employment Hazards: An Investigation of Market Performance.* Cambridge, Mass.: Harvard University Press.

———. 1981. 'Occupational Safety and Health Regulation: Its Impact and Policy Alternatives,' In *Research in Public Policy Analysis and Management,* J. Crecine, ed. Vol. II. Greenwich, Conn: JAI Press.

———. 1983a. 'Alternative Approaches to Valuing the Health Impacts of Accidents: Liability Law and Prospective Evaluations.' *Law and Contemporary Problems* 47(4): 49–68.

———. 1983b. *Risk by Choice: Regulating Health and Safety in the Workplace.* Cambridge, Mass.: Harvard University Press.

———, and Charles O'Connor. 1984. 'Adaptive Responses to Chemical Labeling: Are Workers Bayesian Decision Makers?' *American Economic Review* 74(5).

Weinstein, Milton, Donald Shepard, and Joseph Pliskin. 1980. 'The Economic Value of Changing Mortality Probabilities: A Decision-Theoretic Approach.' *Quarterly Journal of Economics* 94: 375–396.

Zeckhauser, Richard. 1974. 'Risk Spreading and Distribution.' In *Redistribution through Public Choice,* H. Hochman and G. Peterson, eds. New York: Columbia University Press.
———. 1975. 'Procedures for Valuing Lives.' *Public Policy* 23: 419–464.
———, and Donald Shepard. 1976. 'Where Now for Saving Lives?' *Law and Contemporary Problems* 40: 5–45.

Chapter 8

Estimating the Benefits of Environmental Regulations

A. Myrick Freeman III

The purpose of this chapter is to identify and evaluate the available techniques and economic models for estimating other than human health related benefits from environmental regulations. The benefits associated with improved longevity, health, and safety are discussed in the previous chapter. This chapter focuses on the benefits of such things as improved productivity of agriculture and other commercially exploited ecosystems, improved opportunities for recreation, aesthetics, and so forth.

The Structure of a Benefits Estimation Model

The evaluation of any particular technique for estimating the benefits of an environmental regulation involves answering two major questions. First, is the technique based on a conceptually valid theoretical model that captures the principal physical, biological, and economic relationships? Second, are the data requirements of the model reasonable; that is, are the required data available or can they be gathered at reasonable cost? In order to get a clearer picture of the data requirements and the roles of physical, biological, and economic models in benefit estimation, it will be helpful to examine the underlying structure of the benefit estimation problem.

The ultimate objective of the analysis is to measure the welfare gains associated with an environmental regulation. A regulation affects the decisions and behavior of sources of pollution. As a consequence there are changes in the rates at which polluting substances are discharged into the environment. These changes in turn lead to changes in the concentrations of substances in the environment over some geographical area. Changes in measures of environmental quality affect the uses made of the environment by both producers and individuals acting as consumers. It is these changes in the uses of the environment which have welfare implications and which can, in principle, be measured in money terms.

The process by which the benefits of environmental improvement are produced by a regulatory policy has three distinct stages. Estimation of the benefits of a regulation requires knowledge of the relationships involved in each of these stages:

1. Regulatory policy leads to improvements in environmental quality. Emissions standards or effluent charges, if effective, induce polluters

to reduce their discharge of polluting substances. Changes in the temporal and spatial pattern of discharges lead to changes in the temporal and spatial patterns of air and water quality.
2. Changes in environmental quality result in changes in the types and levels of human uses of the environment.
3. Changes in human uses of the environment affect utility or welfare. These changes can be measured by monetary equivalents, measured in terms of willingness to pay.

The first stage is almost entirely noneconomic in nature because it involves a variety of physical, chemical, and biological processes and relationships. The third stage is wholly within the realm of economics because it involves demand and production theory and the theory of economic value. The second stage involves the interface between the noneconomic and economic stages of the production of benefits. Understanding of this stage is essential if empirical estimates of benefits are to be made.

To formalize these relationships, consider the following highly idealized model. Assume there is only one polluting substance discharged into the environment. Let D represent the quantity of this substance which is discharged per year. Suppose that environmental quality can be measured by a single parameter Q. Let X represent the level of an activity which is adversely affected by pollution. Finally let W represent the level of economic welfare associated with X, measured in money. It might be helpful to think of the following concrete example: D is biochemical oxygen demand (BOD); Q is dissolved oxygen (DO); X is recreation days per year. The model can be expressed as:

$$Q = Q(D), (dQ/dD < 0) \quad (1)$$
$$X = X(Q), (dW/dQ > 0) \quad (2)$$
$$W = W(X), (dW/dX > 0) \quad (3)$$

By substitution we have

$$W = f(D), dW/dD < 0 \quad (4)$$

The benefit of a pollution control regulation that reduces D from D_1 to $D_2 (= \Delta D)$ is

$$B = \Delta W = f(D_2) - f(D_1) \quad (5)$$
$$= B(\Delta D)$$

where B is an aggregate of the compensating or equivalent variations of all people affected by the change in D.

Estimating the benefit of a proposed regulation entails first predicting the responses of affected dischargers, that is, ΔD, and then tracing the effects of ΔD through the links described by Eqs. (1)–(3) to calculate the resulting welfare changes in monetary terms. Predicting dischargers' responses to regulations is itself a challenging task that raises issues beyond the scope of

8. The Benefits of Environmental Regulations

this chapter. Suffice it to say that both theory and observation show that the analyst should not assume perfect compliance with the regulation.

This simple representation of the problem obscures a number of details and complications that have to be reckoned with. First, most environmental problems have an important spatial component. Discharges may come from several (or many) sources at different locations. Similarly, environmental quality (Q) may vary across space according to the spatial pattern of discharges, dispersion, and so forth. These spatial characteristics must be reflected in the benefits model. There may be an important temporal dimension to the discharges and measures of environmental quality.

The relationships that relate rates of discharge D to environmental quality Q are very complex, particularly those which describe impacts on the living ecological system. The accuracy of benefits analyses, either in monetary or nonmonetary terms, ultimately rests on the accuracy of this first step. In many cases the overall analysis is segmented into separate parts, the benefits analyst beginning with a relatively fixed set of inputs. Under these circumstances, the final results reflect only the uncertainty in the valuation process and may present a misleading degree of confidence. It is very important to introduce measures of the uncertainty in the overall process in the presentation of the results.

The general relationship $X = X(Q)$ reflects in part the physical and biological consequences of pollution. For example, it could reflect the impact of ozone on agricultural crop yields, of sulfur dioxide on material corrosion rates, or of dissolved oxygen on rates of recreation use of a body of water. Understanding these impacts is the problem of the plant scientist, the materials engineer, or the economist analyzing recreation behavior. These relationships must also incorporate producers' and individuals' behavioral responses to these physical and biological effects. Individuals and producers can engage in averting behavior and mitigating activities. Farmers can shift away from pollution-sensitive crops and change planting times and fertilization regimes. Producers can protect against corrosion by applying coatings or shifting to less sensitive materials. And individuals can shift to substitute recreation sites or alternative recreation activities. Changes in the use of the environment and the welfare implications of these changes depend in part on the opportunities for and the costs of mitigating and averting activities. Benefit estimation models must attempt to capture the major averting and mitigating activities available to people.

The third stage in estimating benefits involves determining the monetary values that people place on such things as increased agricultural productivity and improved opportunities for water-based recreation. Regarding the analysis of this stage, there is a well-developed theory of economic value. The theory provides a number of approaches for estimating these values under different circumstances. The major purpose of this chapter is to review the most important of these approaches in terms of theoretical validity and practical usefulness.

Types of Benefits

Different benefit estimation techniques have characteristics and data requirements that make them more or less well suited for estimating particular types of benefits. In order to organize the discussion of different benefit estimation techniques, it would be helpful to have a classification or taxonomy of benefit categories. Table 8-1 shows in summary form the categories of the benefits of environmental change which are used in this chapter. The first major distinction reflected in this table is that between production or market values and what is called individual values in this work. Production or market values arise in those cases where some attribute of the environment is an argument in the production and cost functions for a marketed good. A beneficial environmental change is reflected in an increase in economic productivity and a reduction in the cost of producing

Table 8–1. Categories of environmental benefits*

I. Production/Market Value: Some attribute of the environment is an argument in the production and cost function for a market good, e.g., forest products, commercial fisheries

II. Individual Values: Some attribute of the environment is an argument in individuals' utility functions
   A. Use Values: Based on *in situ* use of the environment
   B. Nonuse/Intrinsic Values:
      1. Existence values – unrelated to present or future uses by the individuals
      2. Option values – related to uncertainty of future use and/or supply

* Adapted from Desvousges, Smith, and McGivney (1983).

the marketed good. This in turn results in changes in marketed quantities, product prices, factor prices, rents, and profits. Standard economic models based on supply and demand functions can be used to derive measures of benefits from these market changes.

In the case of production and market values, environmental changes affect individuals only by changing the prices of goods they purchase or their incomes. In contrast, in the case of individual values, some attribute of the environment enters directly as an argument in the individuals' utility function. Within the category of individual values, the major distinction is between those values associated with the direct or *in situ* use of the environment and nonuse or intrinsic values.

The *in situ* use of the environment is an activity that absorbs the scarce resources of the individual, including but not limited to time. For example, the individual may have to incur time and other costs to travel to some location to engage in some recreational activity. Under some circumstances it is possible to use information on the observed behavior associated with the use of the environment, such as travel cost, to infer the individual's demand for the services of the environment, thus providing a basis for inferring

8. The Benefits of Environmental Regulations

values. The key here is the link between observable activities – that is, use of the environment – and economic value.

In contrast, this link is not present in the case of nonuse or intrinsic benefits. Nonuse or intrinsic values are defined as those benefits or welfare gains to individuals that arise from environmental changes independently of any direct use of the environment. The term nonuse is preferred to the more common nonuser, since it is possible for a user to realize additional nonuse or intrinsic benefits over and above those associated with his direct use of the environment.

Since use values arise from *in situ* use of the environment, in principle it is possible to exclude individuals from using the environment and hence from experiencing those benefits. Thus one can contemplate the emergence of markets for those environmental services generating use values if property rights were defined and transactions costs were low. In contrast, exclusion is not possible for existence values and probably not practical for option values. These values are public goods; and markets cannot be relied upon to provide the optimal quantities of resources producing intrinsic values.

The category of intrinsic values can be further divided into pure existence values and option values related to uncertainty concerning future demand and/or availability of the ecosystem for possible use. The concept of pure existence value was apparently first suggested by Krutilla (1967) and was further discussed in Krutilla and Fisher (1974: 124). The possible motivations or rationales for existence values include bequest, preservation, and other altruistic motives.

A number of such nonuse or intrinsic benefits are related to ecological attributes. Support for the protection of endangered species and protection of critical habitats for those species represents an intrinsic valuation process. Visibility in pristine areas is also valued by nonusers. Improved visibility may provide use benefits in addition.

Weisbrod (1964) first introduced the term option value into the literature of benefit-cost analysis twenty years ago. He argued that an individual who was unsure of whether he would visit a site such as a national park would be willing to pay a sum in excess of his expected consumer surplus to guarantee that the site would be available should he wish to visit it. Option values were said to arise when an individual was uncertain as to whether he would demand a good in some future period and was faced with uncertainty about the supply or future availability of that good. If option price is defined as the maximum sum the individual would be willing to pay to preserve the option to visit the site before his own demand uncertainty is resolved, then the excess of option price over expected consumer surplus can be called option value. Option value is distinct from a use value in that it arises not from the use of the site itself but from uncertainty over the site's availability or existence to meet possible future demands.

Weisbrod apparently viewed the existence of positive option value as being intuitively obvious. But as subsequent analysis has shown, option value can either be positive or negative depending upon the particular circumstances. A key paper is by Schmalensee (1972). Bishop (1982) has provided a useful review and extension of the literature since Weisbrod's paper. The questions of the sign of option value (whether positive or negative) and use of option value in benefit-cost analysis will be taken up below.

Market Benefits

The symbol Q denotes a factor input whenever Q enters positively in the production function of a market good (or when pollution enters negatively). When Q is a factor of production, changes in Q lead to changes in production costs, which in turn affect the price and quantity of output or the returns to other factor inputs, or both. The benefits of changes in Q can be inferred from changes in these observable market data. There are several examples where Q can be interpreted as a factor input. The quality of river water diverted for irrigation affects the agricultural productivity of irrigated land. The quality of intake water may influence the cost of treating domestic water supplies or the cost of production in industrial operations utilizing water for processing purposes. Agricultural productivity is impaired by some forms of air pollution. And to the extent that air pollution causes materials damages, it can affect the cost of production for a wide variety of goods and services.

Damage function models

Direct cost or damage function models have often been used to estimate the benefits of pollution control. Examples of such studies are given in Freeman (1982). However, these models may give seriously misleading results unless several important simplifying assumptions are in fact valid. For example, suppose an estimate of the relationship between air pollution and the yield of a crop is used to predict the change in output of that crop for a given change in air quality. A damage function model would take the predicted change in output and multiply it by the observed market price of the output to obtain an estimate of the increase in market value of the crop and interpret it as a measure of benefits. However, unless the predicted increase in output is small relative to the total market for this crop, the price of the crop will fall. This, of course, reduces the benefits accruing to producers and, depending upon the price elasticity of demand, may even make producers worse off. And even if the price effect is negligible, the actual increase in output could be substantially less than that predicted by the damage function model. For example, farmers could respond to the increase in productivity by reducing inputs of other costly factors of production or by shifting to other more profitable but pollution-sensitive cultivars or crops.

Similarly, damage function models that value predicted physical damages by the cost of repair or replacement may be in error if producers actually respond to physical damages by materials substitution or changes in input mix. The strength of the cost function and optimization models is that they attempt to capture this wider range of response options either implicitly in the case of the cost function model or explicitly in the case of the simulation model.

Cost function models

The duality or cost function model for estimating market benefits is in principle capable of capturing the benefits accruing through both of these channels. The technique requires the simultaneous estimating of cost or supply functions for affected firms and the demand function for the market output. It is a straightforward matter to calculate changes in producers' and consumers' surpluses for postulated changes in Q. One example of this approach is the study by Mjelde *et al.* (1984), who estimated the benefits to farmers in Illinois of controlling ozone pollution.

A modification of this approach was recently used by Mathtech, Inc. (1982) to estimate the benefits of achieving ambient air quality standards for sulfur dioxide and total suspended particulates. Mathtech estimated cost functions for six 3-digit SIC industries and for the electric utility industry. A pollution measure was statistically significant in five of the seven industry categories examined. Mathtech calculated benefits on the simplifying assumption that lower costs would not result in increases in output. This is equivalent to assuming that the demand function is perfectly inelastic. If in fact lower costs lead to higher outputs and lower prices to consumers, Mathtech's calculations lead to underestimates of benefits.

There are several advantages to the cost function approach. First, it is not necessary to determine damage functions for specific materials. Second, it is not necessary to estimate the stock of each material exposed to pollution. And finally, it is not necessary to determine specific producers' responses to air pollution – such as material substitution and preventive actions. All of these relationships are embodied implicitly in the estimated cost functions. On the other hand, this approach is subject to all of the econometric problems associated with the specification and esitmation of cost and supply functions. And relatively weak but not insignificant pollution effects may be difficult to sort out from the statistical noise in equations which involve a substantial degree of aggregation.

Simulation or optimization models

An alternative approach to estimating market benefits to both consumers and producers involves explicitly modelling the optimization problem of producers. In principle, this approach is capable of taking account of producers' opportunities for changing the mix of outputs and changing input combinations in response to changes in pollution levels. These producer adjustments are captured implicitly in the cost function model, but they can be made explicit and perhaps be represented more accurately in micro or firm level optimization models. One example of this approach is the study of farmers' responses to air pollution by Adams, Thanavibulchai, and Crocker

(1979). They developed and implemented a comprehensive model of producers' and market responses to changes in air quality to estimate the benefits to agriculture of controlling ozone in Southern California. The principal components of their analysis are quadratic programming models for each of four agricultural subregions in California and a set of price forecasting equations for 14 crops. Yield equations giving output per acre for each crop as a function of ozone were taken from the open literature. For a specified set of air quality levels in the four regions, the quadratic programming model was used to compute the profit maximizing allocation of land and other inputs to 14 crops and the outputs of each crop by region. The price forecasting equations were then used to compute producers' profits and consumer surpluses. An improvement in ozone levels would lead to increased outputs and lower prices, thus benefiting consumers. Conceivably, with sufficiently inelastic demand functions, the price decreases could make producers worse off. But in their model, yield increases outweighed price decreases and producers reaped approximately 75 percent of the total benefits of air quality improvement.

The development of reasonably detailed optimization models from scratch for this purpose is a very laborious and time-consuming task. This approach may only be practical when existing programming models can be modified and adapted to the task at hand. The ongoing effort by the National Crop Loss Assessment Network to estimate the benefits of ozone pollution control is an example of this approach based on experimental evidence of the effects of ozone levels on crop productivity and an existing model of farmers' crop mix and input combination decisions (Adams, Hamilton, and McCarl 1984; see also Kopp and Vaughan 1983).

Indirect Methods for Estimating Individual Use Benefits

For an individual who makes some use of an environmental resource or service and whose utility function includes some measure of environmental quality Q as an argument, the benefits to the individual of an improvement in Q are measured by the area under the individual's compensated inverse demand function for Q. Changes in various measures of environmental quality can affect individuals through a variety of channels. This is equivalent to saying there are a variety of alternative specifications of individuals' utility functions and their associated inverse demand functions for Q. In this section several models for benefit estimation based on alternative specifications of individuals' utility functions are identified and their usefulness for estimating different classes of individual use benefits is discussed. For more rigorous development of the theory underlying these methods, see Freeman (1985b) and references therein.

Defensive Expenditures and Averting Behavior

Sometimes an individual can defend himself against the adverse effects of pollution by increasing purchases of some other good (e.g., paint, air conditioning equipment) or by increasing spending on some other activity (e.g., cleaning). The increased spending is a direct cost or a defensive expenditure. Such direct costs or defensive expenditures have been the basis of several estimates of the benefits to households in the form of reduced materials damages and soiling due to air pollution. For discussions of some of these studies see Freeman (1982: Chapters 5 and 8).

Is this approach to benefit estimation logically correct? Can decreases in spending on substitute goods be taken as a measure of benefits of an increase in Q? In general, the answer is no, even in the case when Q and the market good are perfect substitutes. The intuition behind this answer is straightforward. Assume that Q and market good X are perfect substitutes. In this special case, the benefit of a change in Q is equal to the reduction in the spending on X which is required to keep the individual on the original indifference curve. But given a reduction in Q, the individual will not reduce his spending on X so as to stay on the original indifference curve. There is an income effect as well as a substitution effect between X and Q. The increase in Q means that the same level of utility can be maintained with a smaller expenditure on X. As a consequence the individual will reallocate expenditure among all goods, including X, so as to maximize the increase in total utility. This reallocation will result in increases in the expenditure of all goods with positive income elasticities of demand. Therefore, the observed decrease in spending on X will be less than the decrease required to stay on the original indifference curve. And the observed reduction in spending will be an underestimate of the benefits of higher Q.

Similar results have been established rigorously by Courant and Porter (1981). They considered several models of increasing complexity with several kinds of substitute relationships between Q and other goods. They showed that whether changes in defensive spending are good approximations of benefits or seriously over- or underestimate benefits depends on the specific properties of the utility function and the implied relationships between Q and market goods. Their analysis illustrates the importance of developing explicit models of the role of Q in the utility function and how it affects choices of market goods before attempting to draw inferences about the magnitude of the benefits from changes in market goods demands. Simple direct cost and defensive expenditure models may have relatively modest data requirements. But they may lead to serious errors in benefit estimation by their failure to model accurately the technical and behavioral alternatives available to individuals responding to a change in Q.

The Household Production Framework

The household production function model provides a framework for examining the interactions between the demands for market goods and changes in environmental quality. The household production framework is based on the assumption that there is a set of technical relationships among goods used by households in the implicit production of utility yielding final services. Examining the household production technology may be one approach to gaining knowledge of the relationship between demands for market goods and the value of environmental quality change.

One approach to using this information about market goods demands to determine the benefits of changes in Q was outlined by Hori (1975). He showed that under fairly general conditions, knowledge of the household production function, the market goods demands, and the level of Q is sufficient to determine the ratios of the marginal costs of production (i.e., the marginal rate of transformation), MRT_{z_i,z_j}, for any pair of final service flows. If the individual is in equilibrium, knowledge of the z_i values and the marginal costs gives the individual marginal rates of substitution, since $MRS_{z_i,z_j} = MRT_{z_i,z_j}$. Hori argued that this established the marginal rate of substitution as a function of the arguments of the marginal rate of transformation – that is, market goods and their prices and Q. If this is so, integration to the utility function is straightforward.

Whether Hori's approach represents a practical advance in technique over other models may be open to question. Smith (1979) has noted that Hori imposes very restrictive assumptions on the models: knowledge of the whole household production technology is required to solve for the utility function.

There has been at least one attempt to implement a variation on Hori's model based on household production theory. In a recent study for the Environmental Protection Agency, Mathtech (1982) used household expenditure data to estimate a system of demand functions for market goods and the underlying demands for final service flows. They proceded with the following steps:

1. Using a linear expenditure system and assuming weak separability, a system of demand functions was estimated for goods grouped into seven categories: food, shelter, home operations, home furnishings and equipment, clothing, transportation, and health and personal care. Each category was assumed to correspond to one final service flow: for example, the category home operations was to represent the final service flow cleanliness. A measure of air pollution was included as an argument in each of the market goods demand functions.

2. The parameters of the estimated market goods demand functions were used to calculate price or cost and quantity indices for the final services flows. The price or cost indices were increasing functions of air pollution.

3. The price and quantity indices were used to estimate the demand functions for each of the seven final service flows.

4. In those cases where air pollution was statistically significant in market goods demand functions, an improvement in air quality would lower the price of the corresponding final service flow. Calculation of the benefits of improved air quality is straightforward once the corresponding final goods demand functions are known.

The Mathtech study provides evidence of both strengths and limitations in the household production approach. On the plus side this method provides an opportunity within the context of the model to capture the rich variety of possible substitutions between marketed goods and attributes of the environment and among various types of final service flows. In this regard, the model is conceptually superior to the defensive expenditure or direct cost techniques. However, this approach makes severe demands for data and requires sophisticated econometric techniques. Notice that to make use of the available data and to make the estimation problems tractable, the Mathtech study assumed that there were only seven final service flows.

Site-Specific Amenities and the Hedonic Price Model

Where some measure of environmental quality varies systematically across space, and where people desire higher levels of quality at their residence, other things equal, housing prices are likely to be affected by the level of environmental quality. Information on the demand for environmental quality is embedded in the prices of housing in this case. The hedonic price model has been developed as a basis for deriving value information for site-specific amenities from the variation in housing prices as a function of housing attributes including environmental quality (see Chapter 6).

In order to estimate benefits of a nonmarginal change in Q, it is necessary to know the inverse demand function for Q. But for either hedonic wage or property values, the inverse demand function cannot be estimated from data from a single labor or property market unless additional restrictions are imposed. When hedonic wage functions have been used to estimate benefits, the typical approach has been to make some arbitrary assumption about the shape of the inverse demand functions through the known point. See, for example, National Academy of Sciences (1974: 243–255) and Bayless (1982).

Finally, it should be noted in passing that the interpretation of wage differences as reflecting compensation for amenity differences is based on an essentially partial equilibrium view of the economy. There has been relatively little work done on the development of comprehensive general equilibrium models of the economic relations involving production, trade, and labor migration among cities or on the relationships among labor markets, goods markets, land markets, and the generation of pollution externalities. Rosen (1979: 78–79, 84) alludes to some of the problems. Freeman (1979: 118–121) lays out some of the issues in an informal manner. And Roback (1982) and Bartik and Smith (forthcoming) consider interactions between labor markets and the land markets in a general equilibrium interurban setting. These papers are suggestive of some of the problems in interpreting interurban wage differences and in developing and estimating models that fully capture the range of adjustments of both employers and workers to differences in levels of amenities within and between cities. At this stage in its development, the interurban wage model cannot be considered a reliable estimator for the benefits of environmental regulation.

Travel Cost Models and Recreation Benefits

Perhaps as much as one half of the total benefits for controlling water pollution in the United States can be attributed to improvements in water-based recreation opportunities (Freeman 1982: 169–171). An improvement in the quality of water in a given lake or stretch of river may be of no value to an individual unless he engages in recreational activities at that site. If that is the case, then water quality Q and recreation activities measured by visits to the site are weak complements as defined by Maler (1974). This means that if a demand function for visits to the site incorporating Q as an argument can be estimated, then the benefits of an increase in Q can be calculated from the area between the demand curves before and after the improvement in Q, other things being equal.

This by now familiar Hotelling-Clawson-Knetsch travel cost model provides a basis for estimating recreation site demands, as shown in Chapter 6. The basic references are Clawson and Knetsch (1966) and Knetsch (1964). For a clear exposition with a numerical example of its application, see Maler and Wyzga (1976).

There are two major problems in any recreation demand analysis; and how they are dealt with can affect the usefulness of site demand functions for estimating recreation benefits. These problems are (1) the treatment of alternative recreation sites which may be substitutes for the site in question and (2) the role and value of time on site and in travel to the site. These isues are reviewed in Freeman (1979: Chapter 8). More recent references on the treatment of time include Wilman (1980), McConnell and Strand (1981), and Smith, Desvousges, and McGivney (1983).

The most important problem in the application of demand models to benefit estimation is determining the effect of changes in the relevant measure of environmental quality on the site demand curve. Several approaches have been discussed in the literature. One approach involves using time series analysis of the demand at one site when changes in Q have occurred over the relevant time span. Another possibility involves using cross-section analysis of recreation demands at several sites where differences in site qualities help to explain the distribution of visits across sites. These approaches are discussed in Freeman (1979: Chapter 8).

A third approach to introducing quality effects is to estimate demand functions for each site separately without quality variables, and then to attempt to explain differences in the coefficients on price terms by regressing them on the quality variables. To take a simple example, the set of demand funtions

$$V_i = a + b_i P_{V_i}$$

would be estimated by the Hotelling-Clawson-Knetsch technique, where V_i

is visits to the site and P_{V_i} is the travel cost to that site. This equation could include prices of substitution sites. The own-price coefficients would be regressed against the quality variables:

$$b_i = c + dQ_i$$

By substitution, this is equivalent to including interaction terms; that is,

$$V_i = a + cP_{V_i} + dQ_i P_{V_i}$$

This equation should be specified to include only quality variables for site i, or it could include quality variables for other sites as well to test for substitution. For further discussion of this approach and the econometric issues involved, see Saxonhouse (1977). An application of this approach can be found in Desvousges, Smith, and McGivney (1983). This report includes an extensive discussion of the theoretical justification for this model specification.

A fourth approach involves the application of the logit model to data from multiple sites. It may often be the case that when faced with an array of alternative sites at various distances and with different Qs, individuals will choose to make most of their visits to one or two sites and make no visits to most of the alternative sites in the region. If this is the case, estimation of a system of visitation equations by ordinary least squares would be inappropriate because of the large number of zero values for the dependent variable. The ordinary-least-squares specification implies that the effect of a change in Q at site j on visits by an individual to site i is the same regardless of whether site j is visited or not. In these circumstances, the logit model can be used to analyze individual site visitation data. The logit model can be interpreted as estimating the probability that an individual will visit a given site as a function of characteristics of the individual and the available sites. A derivation of the logit model and an application to recreation site visit data for the Boston region is presented in Feenberg and Mills (1980).

The logit model provides a straightforward basis for calculating the benefit per visit of an improvement in Q at a site, at least if a measure of price or travel cost is included in the logit equation. Other things equal, an increase in Q leads to an increase in the probability of visiting that site and a higher level of utility for the individual. Total differentiation of the logit equation makes it possible to calculate the increase in price or travel cost per visit which would leave the probability of visiting unchanged after an increase in Q. This compensating increase in price is the benefit per visit of the higher Q. Feenberg and Mills provide sample calculations for their Boston model (1980: 115). They also derive a measure of total benefits which takes account of the likelihood that the total number of visits to all sites will increase with improvements in Q at one or more sites.

The last approach to estimating the value of Q in recreation to be considered here involves a combination of aspects of the hedonic and travel cost models. It has been developed and applied to estimate the value of fishing success by Brown and Mendelsohn (1983) and has been termed by them the hedonic travel cost model. In their model a recreation site is a differentiated good which can be described by a vector of its attributes or qualities. An individual at a given location faces an array of alternative sites with different characteristics. Each is available at a different price, where this includes any entry fee and the cost of travel to the site. A hedonic price function can be estimated for these sites as a function of site attributes. This hedonic price function is specific to the individual's location, since a major cause of the variation in prices of sites is variation in the travel cost to those sites from the individual's location.

If there were a sufficient number of sites so that the availability of different attributes could be represented by continuous functions, then utility-maximizing individuals would select sites so that their marginal willingness to pay for each attribute would be equal to its marginal implicit price. With knowledge of sites actually selected by individuals, it would be possible to calculate the marginal willingness to pay or marginal benefit of improvements in any of the site attributes. But as was pointed out in the discussion on hedonic price models (see Chapter 6), in the absence of additional information or added restrictions it is not possible to estimate demand functions for attributes for this data.

Brown and Mendelsohn (1983) solved this problem in the recreation setting by estimating separate hedonic price functions for groups of individuals residing in different localities within a region by making use of the range of recreation sites in that region. Their specific application involved fishermen in various residential localities and river fishing sites in western Washington. In effect, they treated each locality as a separate market for characteristics for purposes of hedonic analysis. After calculating hedonic price functions for each locality and determining the observed equilibrium marginal implicit prices of attributes, they estimated inverse demand functions for attributes by regressing observed marginal implicit prices for all fishermen against observed quantities of attributes and socioeconomic variables.

Nonuse/Intrinsic Benefits

As discussed in an earlier section, intrinsic values can be categorized as either (1) pure existence values unrelated to any present or future use of the environmental resource or (2) option values that arise from actions to reduce uncertainty concerning *future* supply of the resource for those who are actual or potential users. Option value is considered a nonuse value since it is unrelated to *present* use. In this section approaches to measuring each of these types of values are discussed in turn.

Pure existence or preservation value may be significant for policies to prevent irreversible alteration or destruction of a unique scenic resource or ecological system or for the protection of an endangered species. People who presently make no use of such an environmental resource may still be willing to pay for the preservation of the resource. Several authors have discussed possible motivations for such pure existence values (e.g., Krutilla 1967, McConnell 1983, and Randall and Stoll 1983). Existence values could also be present for users. For example, an individual might be willing to pay to preserve a critical habitat even if he knew that part of the effort to preserve the habitat would entail barring human use of it.

Existence value is a pure collective or public good. Since it is unrelated to present use of the environmental resource, there are no indirect methods of measuring such values. Although individuals may make financial contributions to environmental organizations and contribute their time to activities designed to assure the preservation of environmental resources, these activities can only be interpreted as supporting the hypothesis that preservation values are positive. Because of free rider problems, because environmental organizations provide a variety of bundled preservation and other services to contributors, and because these institutions do not have a mechanism for extracting the *maximum* willingness to pay, there are at present no models for using such revealed behavior to infer existence or preservation benefits. It appears that contingent valuation methods provide the only feasible technique for estimating pure existence value. The contingent valuation method will be discussed in the next section.

The nature of option value might best be illustrated with the following example. Consider a national park containing unique scenic and ecological resources. Suppose that at the beginning of each year all individuals are uncertain *ex ante* as to whether they will wish to visit the park during that year. But during the course of the year, some people will in fact visit the park. Indirect benefit estimation techniques can be used to measure the value of the park to those who actually visit. The consumer surpluses of the actual users represent an *ex post* measure of the value of the park to actual users. This value can also be interpreted as the expected value of the consumer surpluses of the population as a whole.

Alternatively, we could ask each individual at the beginning of the year for his maximum willingness to pay to preserve the right or option to visit the park during that year. This *ex ante* payment is the option price. The early literature on option value was concerned with the question of whether the aggregate option price would be greater that the aggregate expected consumer surplus. If option price was greater, the excess was termed option value.

More recent theoretical analysis has identified three sets of issues which collectively make the role of option value in benefit-cost analysis much more complicated than the preceding example suggests. First, it has been shown that in the context of the preceding example, option price can be either greater or less than the expected consumer surplus depending on the nature and source of uncertainty concerning demand (Schmalensee 1972; Hartman and Plummer 1982). Furthermore, the difference between option price and expected surplus (whether positive or negative) might be trivially small or quite large also depending on the circumstances (Freeman 1984). This raises a question of whether something which can be either positive or negative should be termed a "value."

The second question is what is the appropriate measure of welfare change where uncertainty is present: option value, expected surplus, or something else. Ulph (1982) has shown that some fundamental questions of social welfare criteria are involved. And even in a Paretian context, Graham (1981) has shown that the appropriate welfare measure depends on the institutions available for individuals to diversify risks and for collecting payments in different states of nature.

The third question involves the nature of uncertainty of supply or availability of the environmental resource to potential users. The early option value models assumed that if the individual paid the option price, the resource would be available with certainty, but if the individual did not pay his option price, he would be excluded from the resource. But other conditions of supply uncertainty are possible. For example, there may be a nonzero probability of availability in the absence of the project or the absence of payment of option price by the individual. And policies may only reduce the uncertainty of supply, not eliminate it. It has been shown that even in the case of certainty of demand on the part of the individual, the relationship between option price and the expected surplus of the project is ambiguous, that is, such supply side option values can be either positive or negative (Freeman 1985a).

This discussion has shown that the conventional view of option value as a separate category of nonuse benefits is too simplistic. Option value can be either positive or negative; and the algebraic sum of option value and expected surplus may not be the appropriate welfare measure, in any case. But suppose that option price is the desired welfare measure. How can it be estimated? Probably the best approach is to ask people their maximum willingness to pay – that is, to use the contingent valuation method. Examples include Desvousges, Smith, and McGivney (1983) and Brookshire, Eubanks, and Randall (1983). Alternatively, if estimates of expected surplus are available, it might be possible to use an analytical model to compute option price as a function of expected surplus, the relevant probabilities, and some measure of risk aversion (see Freeman 1984).

Contingent Choice

The contingent valuation method (CVM) for estimating benefits involves asking people how much they would be willing to pay for a specified environmental improvement. The CVM is just one of a family of contingent choice methods (see Chapter 6). Contingent choice techniques for estimating benefits involve asking people to place themselves in a hypothetical situation designed by the investigator and to respond to specific questions such as how much they would be willing to pay for a specified change, which of several alternatives they would choose, or what activities they would undertake.

There are many questions concerning the accuracy of contingent choice valuations, especially for environmental resources that are difficult to value such as preservation and existence values and the protection of endangered species. Accuracy means the degree of correspondence between an individual's stated value and his true value; and true value means the value that would be revealed by an individual with full knowledge of the characteristics of the environmental resource being valued and full knowledge of his own preferences in the relevant region. The author has argued in the past that contingent choice methods contained relatively weak incentives for accurate responses (Freeman 1979: Chapter 5).

A more immediate problem with many contingent choice studies may be that they ask people to place themselves in unfamiliar portions of their preference functions by asking them to place values on environmental changes with which they have little or no experience. For elaboration of these points see the contribution of Freeman in Cummings, Brookshire, and Schulze (forthcoming). It appears that the contingent choice method is likely to work best for those kinds of problems where it is needed least; that is, where respondents' experiences with changes in the level of the environmental good have left a record of tradeoffs, substitutions, and so forth, which can be the basis of indirect estimation of value. But for those problems which need something like the CVM most – that is, where individuals have little or no experience with different levels of the environmental good – the CVM appears to be least reliable.

Nevertheless, for such things as ecological and existence benefits, indirect methods of benefit estimations are not available. So the CVM, despite its problems, is still likely to be used in these cases.

Bowdoin College

References

Adams, Richard M., Scott A. Hamilton, and Bruce A. McCarl. 1984. 'The Economic Effects of Ozone in Agriculture.' Unpublished.
———, Narongsakdi Thanavibulchai, and Thomas D. Crocker. 1979. *Methods Development for Assessing Air Pollution Damages for Selected Crops within Southern California.* Washington, D.C.: U.S. Environmental Protection Agency.
Bartik, Timothy J. and V. Kerry Smith. Forthcoming. 'Urban Amenities and Public Policy.' In E. S. Mills, ed., *Handbook on Urban Economics.* Amsterdam: North Holland.
Bayless, Mark. 1982. 'Measuring the Benefits of Air Quality Improvements: A Hedonic Salary Approach.' *Journal of Environmental Economics and Management* 9: 81–99.
Bishop, Richard C. 1982. 'Option Value: An Exposition and Extension.' *Land Economics* 58: 1–15.
Brookshire, David S., Larry S. Eubanks, and Alan Randall. 1983. 'Estimating Option Prices and Existence Value for Wildlife Resources.' *Land Economics* 59: 1–15.
Brown, Gardner M. and Robert Mendelsohn. 1983. 'The Hedonic Travel Cost Method.' Unpublished.

Clawson, Marion and Jack l. Knetsch. 1966. *Economics of Outdoor Recreation*. Baltimore: Johns Hopkins University Press.

Courant, Paul N. and Richard Porter. 1981. 'Averting Expenditure and the Cost of Pollution.' *Journal of Environmental Economics and Management* 8: 321–329.

Cummings, Ronald, David S. Brookshire, and William D. Schulze. Forthcoming. *Valuing Environmental Improvements: A State of the Arts Assessment of the Economic Aspects of the Contingent Valuation Method*.

Desvousges, William H., V. Kerry Smith, and Matthew P. McGivney. 1983. *A Comparison of Alternative Approaches for Estimating Recreation and Related Benefits of Water Quality Improvements*. Research Triangle Institute.

Feenberg, Daniel and Edwin S. Mills. 1980. *Measuring the Benefits of Water Pollution Abatement*. New York: Academic Press.

Freeman, A. Myrick III. 1979. *The Benefits of Environmental Improvement: Theory and Practice*. Baltimore: Johns Hopkins University Press.

———. 1982. *Air and Water Pollution Control: A Benefit Cost Assessment*. New York: Wiley.

———. 1984. 'The Sign and Size of Option Value.' *Land Economics* 60: 1–13.

———. 1985a. 'Supply Uncertainty, Option Price, and Option Value.' *Land Economics* 61: 176–181.

———. 1985b. 'Methods for Assessing the Benefits of Environmental Programs.' In Allen V. Kneese and James Sweeney, eds., *Handbook of Environmental and Resource Economics*. Amsterdam: North Holland.

Graham, Daniel A. 1981. 'Cost-Benefit Analysis under Uncertainty.' *American Economic Review* 71: 715–725.

Hartman, Richard and Mark Plumer. 1982. 'Option Value under Income and Price Uncertainty.' Unpublished. University of Washington.

Hori, H. 1975. 'Revealed Preference for Public Goods.' *American Economic Review* 65: 197–991.

Knetsch, Jack L. 1964. 'Economics of Including Recreation as a Purpose of Eastern Water Project.' *Journal of Farm Economics* 46: 1148–1157.

Kopp, Raymond J. and William J. Vaughan. 1983. 'Agricultural Sector Benefits Analysis for Ozone: Methods Evaluation and Demonstration.' Unpublished.

Krutilla, John V. 1967. 'Conservation Reconsidered.' *American Economic Review* 57: 777–786.

——— and Anthony C. Fisher. 1974. *The Economics of Natural Environments: Studies in the Valuation of Commodity and Amenity Resources*. Baltimore: Johns Hopkins University Press.

Maler, Karl-Goran. 1974. *Environmental Economics: A Theoretical Inquiry*. Baltimore: Johns Hopkins University Press.

——— and Ronald Wyzga. 1976. *Economic Measurement of Environmental Damage*. Paris: Organization for Economic Cooperation and Development.

Mathtech. 1982. *Benefits Analysis of Alternative Secondary National Ambient Air Quality Standards for Sulfur Dioxide and Total Suspended Particulates*. Princeton: Mathtech.

McConnell, Kenneth E. 1983. 'Existence and Bequest Value.' In Robert D. Rowe and Lauraine G. Chestnut, eds., *Managing Air Quality and Scenic Resources at National Parks and Wilderness Areas*. Boulder: Westview.

——— and Ivar Strand. 1981. 'Measuring the Cost of Time in Recreaction Demand Analysis: An Application to Sport Fishing.' *American Journal of Agricultural Economics* 63: 153–156.

Mjelde, James W., Richard M. Adams, Bruce L. Dixon, and Philip Garcia. 1984. 'Using Farmers' Actions to Measure Crop Loss Due to Air Pollution.' Unpublished.

National Academy of Sciences, Coordinating Committee on Air Quality Studies. 1974. *Air Quality and Automobile Emission Control. Vol. 4, The Costs and Benefits of Automobile Emission Control*. Washington, D.C.: National Academy of Sciences.

Randall, Alan and John R. Stoll. 1983. 'Existence Value in a Total Valuation Framework.' In Robert D. Rowe and Lauraine G. Chestnut, eds., *Managing Air Quality and Scenic Resources at National Parks and Wilderness Areas*. Boulder: Westview.

Roback, Jennifer. 1982. 'Wages, Rents, and the Quality of Life.' *Journal of Political Economy* 90: 1257–1278.

Rosen, Sherwin. 1979. 'Wage-Based Indices of Urban Quality of Life.' In Peter Mieszkowski and Mahlon Straszheim, eds., *Current Issues in Urban Economics*. Baltimore: Johns Hopkins University Press.

Saxonhouse, Gary R. 1977. 'Regressions from Samples Having Different Characteristics.' *Review of Economics and Statistics* 59: 234–237.

Schmalensee, Richard. 1972. 'Option Demand and Consumer Surplus: Valuing Price Changes Under Uncertainty.' *American Economic Review* 62: 813–824.

Smith, V. Kerry. 1979. 'Indirect Revelation of the Demand for Public Goods: An Overview and Critique.' *Scottish Journal of Political Economy* 26: 183–184.

———, William H. Desvousges, and Matthew P. McGivney. 1983. 'The Opportunity Cost of Travel Time and Recreation Demand Models.' *Land Economics* 59: 259–278.

Ulph, Alistair, 1982. 'The Role of Ex Ante and Ex Post Decisions in the Valuation of Life.' *Journal of Public Economics* 18: 265–276.

Weisbrod, Burton A. 1974. 'Collective Consumption Services of Individual Consumption Goods.' *Quarterly Journal of Economics* 77: 71–77.

Wilman, Elizabeth A. 1980. 'The Value of Time in Recreation Benefits Studies.' *Journal of Environmental Economics and Management* 7: 272–286.

Reviewers

The following persons reviewed all or portions of this volume. The editors and authors gratefully acknowledge this contribution but assume all responsibility for the contents of the volume.

Arnold Aspelin
　Enviromental Protection Agency

Robert C. Barnard
　Cleary, Gottlieb, Steen & Hamilton

Nathaniel F. Barr
　Department of Energy

Clyde J. Behney
　Office of Technology Assessment, U.S. Congress

Edwin L. Behrens
　Proctor & Gamble Company

Tayler H. Bingham
　Research Triangle Institute

Gale A. Boyd
　Argonne National Laboratory

Larry Braslow
　Occupational Safety and Health Administration

Robert Brown
　Food and Drug Administration

Shepard C. Buchanan
　Bonneville Power Administration

Alan Carlin
　Environmental Protection Agency

Richard T. Carson
　Resources for the Future

Margriet F. Caswell
　University of California

Lauraine Chestnut
　Energy and Resource Consultants, Inc.

Paul Cho
　Department of Energy

Coralyn Colladay
　Department of Health and Human Services

Robert Cooper
　Office of the Secretary of Defense

Stephen W. Cowdin
　California Department of Water Resources

Paul F. Deisler, Jr.
　Shell Oil Company

William H. Desvousges
　Research Triangle Institute

Roger Dower
　Environmental Law Institute

Erica Drazen
　Arthur D. Little, Inc.

Steven F. Edwards
　University of Rhode Island

Neil Eisner
　Department of Transportation

Donna Fishman
　Department of Health and Human Services

Roy Fox
　Bonneville Power Administration

Art Fraas
　Office of Management and Budget

Anthony E. Goldin
　Occupational Safety and Health Administration

Marcia Gowen
　The World Bank/East-West Center

Robert B. Grant
　Department of Commerce

Reviewers

Robin Gregory
  Decision Research
W. Michael Hanemann
  University of California
Peter Barton Hutt
  Covington & Burling
James P. Kahan
  Rand Corporation
Steve Kasower
  California Department of Water Resources
David A. King
  University of Arizona
Karl Kronebusch
  Office of Technology Assessment, U.S. Congress
J. Steven Landefeld
  Department of Commerce
Allan Lewis
  Federal Aviation Administration
R. A. Libby
  Norwich Eaton Pharmaceuticals, Inc.
Ralph Luken
  Environmental Protection Agency
Douglas MacLean
  University of Maryland
Brian Mannix
  Office of Management and Budget
Gail H. Marcus
  Congressional Research Service, U.S. Congress
Wendy Max
  California State University
Al McGartland
  Environmental Protection Agency
Ron McHugh
  Environmental Protection Agency
Walter Milon
  Environmental Protection Agency
Robert C. Mitchell
  Resources for the Future
John Morrell
  Office of Management and Budget
Nick Nichols
  Environmental Protection Agency
Timothy O'Leary
  Chemical Manufacturers Association
Rufus Pepper
  Department of Interior
Judith Pitcher
  Consumer Product Safety Commission
Raymond Prince
  James Madison University
Warren J. Prunella
  Consumer Product Safety Commission
Jerome S. Puskin
  Nuclear Regulatory Commission
Gregory Rodgers
  Consumer Product Safety Commission
Fred Siskind
  Department of Labor
Katherine G. Sommers
  National Academy of Sciences
Chum Lee Soon
  Simon Fraser University
Norman Starler
  Bureau of Reclamation
Robert Stavins
  Harvard University
Joe B. Stevens
  Oregon State University
John R. Stoll
  Texas A&M University
Samuel Stumpf
  Vanderbilt University
Carlos Tekez
  Office of Management and Budget
Chris Whipple
  Electric Power Research Institute
Zach Willey
  Environmental Defense Fund
H. R. Woltman
  Food and Drug Administration
Ronald E. Wyzga
  Electric Power Research Institute
Bill Zamule
  Consumer Product Safety Commission

Index

aggregation, additive versus multiplicative models, 129–131, 134, 139–140; of attribute levels (in econometric approach), 93–95; of benefits, 15–16, 21, 23–24, 43, 54, 63–65, 97, 152–154; of component utilities (in multi-attribute approach), 127–133, 136, 152; of data, 90, of individual preferences, 63–65, 95–96, 147; of utilities, 127–141; of utility and value functions, 146–149; of welfare changes across individuals, 135–142
air quality, 10, 11, 22, 217–218, 220
automobile safety (including airbags and seatbelts), 35, 36-37, 38, 46; subsidies for fatalities, 39

Bayesian confidence interval, 60; Bayesian group decision making, 137
behavior, individual, 66–68; societal, 68–70; informed consent, 69–70; see also risk
benefits, the concept of, 16–21, 52–53; evaluating for regulation, 10, 20–21, 36–39, 55–65, 93, 211–227; *ex ante* versus *ex post*, 149–150, 225; qualitative and quantified, 8–9
benefit-cost analysis (BCA), 4–9, 13–34, 43, 196; compared with other approaches, 27–31; critics of, 18–21; and E. O. 12291, 13, 31, 42–43

centralized approach, 104
Consumer Product Safety Commission (CPSC) and identification of target population, 92; contingent ranking method (CRM), 166–169, 180; contingent valuation method (CVM), 10–11, 19, 20, 153–166, 168, 169, 180, 182, 204–205, 226–227
costs, concept of, 21–23; *ex ante* and *ex post*, 22; econometric model, 22; engineering method, 22–23; industry-supplied, 45; opportunity, 21, 24, 25, 56
cost-effectiveness analysis (CEA), 6, 23, 205–207; compared with other approaches, 27–28

decentralized approach, 103–104
decision-making, process, 27, 37–38, 42–48, 53–65, 69–70, 76, 102–103, 105–106; frameworks compared, 27–31; sequential, 105–106; technical methods compared with normative approaches, 27–31; *see also* process approach
Department of Transportation (DOT), 35, 43
disasters, natural, 35, 38; and Federal Flood Protection Act, 37
discount rate, 24–26, 58, 146–147, 151, 196, 204
double counting, 11, 59

ecological issues, 10, 57, 215
econometric (revealed-preference) approach, 22, 92–102, 107–111, 155–156; attribute levels, 93, 94–95; implicit market approach, 96–97, 108–109
economic impact analysis, compared with other approaches, 30
efficiency approach. *See* Pareto efficiency
engineering cost estimates, 22–23
environment, and contingent valuation methods, 204–205; evaluating (non-health-related) benefits of regulations, 211–227; impact statement, compared with other approaches, 29; identifying benefits, 9–11, 21, 31, 111; National Environmental Policy Act, 29; New Source Performance Standards, 42
Environmental Protection Agency (EPA), 10–11, 207, 220; alternatives to regulation, 38–39

Index

equity, 63–65, 140, 193; weights, 138, 141–142, 147
Executive Order 12044, 1–2
Executive Order 12291, 2–5, 13, 31, 40–41, 52–53; five-step procedure in applying, 41–48
expressed preferences. *See under* preferences
externalities, 70, 86, 94, 122

hazards, excessive, and OSHA, 64; hazardous industries, 43, 66; hazardous materials, 35–36, 47, 78. *See also* risk
health, and human capital approach, 198; and multiattribute approach, 125–126, 129, 130–132; valuation of, 6–7, 10, 111, 193–210; muiltiple impacts of policies, 204–206
hedonic price method (HPM), 10–11, 171, 176, 181, 199; and site-specific amenities, 221–222, 224; hedonic price function, 174–176
Hicksian measures, 17–18. *See also* Kaldor-Hicks compensation principle
holistic assessment of preferences, 98, 104, 109–110, 118–119, 121
Hotelling-Clawson-Knetsch travel cost model, 222–223
household production function method (HPFM), 179–180, 219–221
housing market, as example of hedonic price method, 173–176
human capital approach in health risks, 198

independence in multiattribute approach, 123–124, 128–129; additive independence (AI), 123–124, 128; difference independence (DI), or weakly difference independence (WDI), 123–124, 128, 133; mutual preferential independence (MPI), 123–124, 128–129, 133, 143; utility independence (UI), 123–124, 129, 133
interview approach. *See* contingent valuation method *and* preferences, revealed intrinsic (or nonuse) benefits or values, 214–215, 224–226

Kaldor-Hicks compensation principle (also known as potential Pareto improvement), 15, 17–18, 23, 63–64

lotteries in utility theory, 114–115, 116, 118, 195

market approach (private goods), 14, 16–18, 23–26, 57, 74–76, 96–102, 104, 111, 150, 161, 199–202, 214–221; implicit, 96–97, 108–109
Marshallian demand function, 16–18
multiattribute assessment of preferences, 104, 109–110, 119–134; multiattribute health benefits, 95
multiattribute utility (MAU) theory, 126, 133–134, 154; algebraic approach, 120, 130–131; independence conditions, 123–125, 128–129; statistical approach, 120, 130–131
multiple objective programming, compared with other approaches, 30, 31

nonmarket approach (public goods), 14, 18–21, 25, 57, 87, 98–100, 161–182, 202–206; alternative approaches to benefits assessment for, 107–111; direct methods of evaluation (CVM, CRM, PM), 107, 162, 163–171, 181; indirect methods of evaluation (HPM, TCM, HPFM, SSM), 107, 171–181
nonuse (or intrinsic) benefits or values, 166, 214–215, 224–226
nuclear power, 36, 37, 43, 47, 59
Nuclear Regulatory Commission (NRC), 36, 41, 44

Occupational Safety and Health Agency (OSHA), 64; chemical labeling regulation as an example of valuation of life and health, 205–207, 208; identifying target population of regulations, 91–92
Office of Management and Budget (OMB), 1–5, 41, 196, 205, 207; *see also* Executive Order 12291
Office of Technology Assessment (OTA), bibliography of benefit-cost analysis and cost-effectiveness analysis, 6
opportunity costs, 21, 24, 25, 56
option value, 215–216, 224–226

Pareto efficiency, compared with benefit-cost analysis, 14–16; of social utility function, 103, 140, 154; potential Pareto improvement (PPI) criterion (also known as Kaldor-Hicks principle), 15, 54, 63–64, 79, 100; Pareto optimality, 54; Pareto superiority, 105–106
paternalistic regulation, 57, 63

Index

perception of risk, 37–38, 46–47; underestimation, 57, 66
petition method (PM), 169–171, 181, 182
planning horizons, 144
preferences, expressed, 70–73, 87–92, 107–110, 204–205; sample selection of, 88–92; revealed (econometric approach), 72, 74–76, 107, 110; socially desirable and strategic responses, 72, 73, 86; misrepresentation, 73, 205; individual, in evaluating benefits, 66–70, 86–92, 95–96, 101, 107; multiattribute assessment of, compared with holistic assessment of, 109–110, 118–119, 119–134; numerical measurement, by direct assessment, 113, 118–119, by hypothetical lotteries, 114–115, 116, 118, 195, by preference differences, 113–114, 118; and subjective expected utility theory, 116–117
present value criterion, 24, 58
private goods. *See* market approach
process approach, 27, 34–48, 86–87, 105–106
public goods. *See* nonmarket approach
public policy, and benefits assessment, 1–12, 23; decision-making frameworks compared, 27–31. *See* regulatory policy

rationality, assumption of, 45–46, 51, 64, 66, 68; and subjective expected utility theory, 116–117, 121, 195
recreation, 10, 214–215; and travel cost methods, 222–224
regulatory agencies, 1–6, 26, 39–41, 47, 204; criticism of, 40; effect of inaction, 56; guidance of, 52–53; identifying options, 54–56; regulatory impact analyses (RIA), OMB guidelines, 2–4, 41, 43. *See also* regulatory policy *and passim*
regulatory policy, alternatives to regulation, 38–39; benefits assessment in, 35–84, 85–159, 211–227; goals of 27, 31, 39; historical overview of, 1–4; identifying policy options, 53–56, 85; identifying and estimating consequences of, 54, 54–61, 85; paternalistic regulations, 57, 63; reasons for regulating, 39–40; *see also* Executive Orders 12044 *and* 12291 *and passim*
resource allocation, 13, 14, 19, 21
revealed preferences. *See under* preferences
risk, acceptable, 9; attitudes toward, 26, 52, 59, 145–146; aversion, 25–26; risk-dollar tradeoff, 194–195, 197–201, 204, 207; information and misinformation about, 46–47, 66, 69–70; low probability-large consequence, 8–9, 38, 145–146; in multiattribute approach, 131–132; perception of, 37–38, 46–47, 66; sharing, 147–149; valuation of, 193–210. *See also* regulatory policy *and* safety risk-benefit analysis, compared with other approaches, 29
risk-risk analysis, compared with other approaches, 27–28

safety, attitudes toward, 46–47, 59, 197; evaluation of, 60, 75, 111, 132, 197; automobile safety, 35, 36–37, 38, 39, 46
sampling. *See under* preferences
scientists, and estimations for regulation, 43, 60–61
seatbelts. *See* automobile safety
sensitivity analyses, 65, 97–98, 142, 205–206
shadow prices, 25, 26, 76
site substitution method (SSM), 179
social utility approach, 64, 86, 102, 111–156; overview and summary of, 151–156. *See also* social utility function *and* utility functions
social utility (or welfare) function, 135–149; coherence approaches, 137–138; efficiency approaches, 140–141; ethical approaches, 138–139; for time streams of consequences, 147–149. *See also* utility functions
stakeholders (interested parties), 36, 37, 43, 44–48, 88–92, 152. *See also* preferences
strategic bias (strategic misrepresentation of preferences), 86, 101, 121, 164–165
subjective expected benefit (SEB), 103–104
subjective expected utility (SEU), 116–117, 145, 195
survey method. *See* contingent valuation method

technology-based standards, 21–22; compared with other approaches, 27–28
time preferences, 23–25, 58, 79, 145–146
time, risk, and uncertainty, in benefit-cost analysis, 23–25, 102–106, 142–151; in social utility model, 142–151; intertemporal utility functions, 144–146, 147–151
time streams of consequences, 24, 103, 104, 142–150, 153

tradeoffs, 99; risk-dollar, 194–195, 197–201, 204, 207; in social utility approach, 112–115, 131–133
travel cost method (TCM), 176–178, 179, 181, 182; and recreation benefits, 222–224

unanimous indifference implies group indifference (UIIGI principle, 80, 137–139
uncertainty, in aggregation, 65; in assessments, 11, 25–26, 59–65, 93; in benefit-cost analysis, 23–26; in econometric approach, 93, 94, 102–106; in subjective expected utility theory, 117, 121
utility functions, 45–46, 116, 118, 119–120, 124–127, 135–149, 218. *See also* value functions, social utility function, *and* social utility approach
utility theory, 115–117. *See also* social utility approach

value functions, 116, 118, 124–125, 143–149. *See also* utility functions
value of life, 194–198, 200–202, 207; quality, 196–197; quantity, 93, 196, 200–202; heterogeneity, 197–198, 200–201; human capital approach to, 198

wealth, and attitude to risk, 194, 197
welfare economics, adapted to benefit-cost analysis, 16–18, 32, 45–46. *See* social utility approach *and* social utility function
willingness to accept (WTA), 54, 62, 100–102, 131–132, 164–167
willingness to pay (WTP), 16, 24, 54, 62, 63–64, 100–109, 127, 131–132, 164–167, 170; and environment, 212, 215, 225, 226; in risks to life and health, 193–196, 198, 205, 207
workers' compensation, 202–203